CONTENTS

interstellar migration and the human experience

interstellar migration and the human experience

Edited by Ben R. Finney and Eric M. Jones

University of California Press
Berkeley · Los Angeles · London

University of California Press
Berkeley and Los Angeles, California

University of California Press, Ltd.
London, England

First Paperback Printing 1986
ISBN 0-520-05898-4

Library of Congress Cataloging in Publication Data

Interstellar migration and the human experience

Proceedings of the Conference on
Interstellar Migration held at
Los Alamos in May 1983

Bibliography: p.
Includes index.

1. Outer space—Exploration-Congresses.
2. Interplanetary voyages—Congresses.
3. Space colonies—Congresses.
I. Finney, Ben R.   II. Jones, Eric M.
III. Los Alamos conference on
Interstellar Migration (1983:
Los Alamos National Laboratory)

TL787.158   1985   303.4'83   84-16282

Printed in the United States of America

1 2 3 4 5 6 7 8 9

Contents

Yet another book about space? Yes, but with a difference, for this one explores the human as well as the technological side of our future beyond Earth.

Usually, space means only rockets, computers, alien concepts, and exotic worlds. But we maintain that space also has a very human meaning. The Space Age is maturing. It is now more than a quarter century since Sputnik. We have visited the Moon, sent spaceships to reconnoiter distant planets, and now orbit scientific laboratories around Earth. In the not too distant future there will be large manned space stations and the first outposts on the Moon and Mars. Then as we learn to do things more efficiently and cheaply, permanent settlements of men, women, and children will be established in space. As has happened so many times in our past, more ordinary folk concerned with making a living and raising their families will follow in the wake of the explorers.

Migration into space may be a revolutionary step for humanity, but it is one that represents a continuity with our past. Although the technologies that make space settlement a real possibility seem new and strange to us, the process of using technology to expand into new areas is not. The story of the rise of humanity—stretching back some five million years to a time when those most adventurous of apes left the shelter of the African forests to walk upright in the grasslands—is a story of cultural as well as physical evolution. Our ancestors learned how to use technology, both material and social, to expand into new regions so that now our species thrives in virtually all of Earth's environments. This long record of expansion and adaptation, with both successes and failures, has much to teach us about our prospects in space—in the near term as we leave the terrestrial cradle and learn the basis of living in space and in the long term as we leave the Solar System for the stars.

This book constitutes the proceedings of a Conference on Interstellar Migration held at Los Alamos in May 1983, to which we invited biologists, humanists, social scientists, and physical scientists to consider the human future in space. This attempt to bridge the many gaps between disciplines came naturally to us, for as an anthropologist (Finney) and an astrophysicist (Jones) we had long realized that so portentous a subject as space migration cried out for attention by all the disciplines that treat the human condition in addition to those more traditionally involved in space science. Hence, for the conference we brought together anthropologists, demographers, historians, paleontologists, and philosophers as well as astronomers, physicists, and machine intelligence specialists. To say that the conference was a learning experience is an understatement. Some physical scientists came believing perhaps a little too much in the power of high technology for the comfort of the social scientists and humanists, whereas some of the latter were skeptical of either the feasibility or the advisability of settling space. ("You think that anything not forbidden by physics will happen," complained one.) But by the end of a very intensive week, we had all learned something about the human experience as a migratory species and about how resources in space might be used to support self-sufficient human communities there.

As editors we have encouraged the authors to retain their special viewpoints. Because the following papers cover such a wide range of topics, we have grouped them into sections and have added introductory commentary to each in an attempt to convey a coherent story. As contributors we too have our own viewpoints; we hope that our editorial comments will help to communicate the lively individualism that characterized the conference, highlighting both disagreement and consensus. Two of our papers published here were not presented orally at the conference. "The Exploring Animal" is an adaptation of a paper, "From Africa to the Stars: the Evolution of the Exploring Animal," presented at the Fifth Conference on Space Manufacturing held at Princeton University two weeks before our conference and published in *Space Manufacturing 1983* (James D. Burke and April S. Whitt, eds., American Astronautical Society, San Diego, 1983, pp. 85-103). "The Prince and the Eunuch" was presented in another form at the annual meeting of the American Association of Aeronautics and Astronautics, held at Baltimore in May 1982. In addition, the paper by William Hodges in Section II was written after the conference in response to Kenneth Wachter's contribution.

The Los Alamos Conference on Interstellar Migrations was supported by the Los Alamos National Laboratory and its Institute of Geophysics and Planetary Physics, the California Space Institute, the Planetary Society, and the School of American Research. We thank these institutions for their financial support, and we also acknowledge the enthusiastic support of the following individuals: P. Coleman (Los Alamos/IGPP), D. Kerr (Los Alamos), M. Birely (Los Alamos), C. Sagan (Planetary Society/Cornell University), L. Friedmann (Planetary Society), D. Schwartz (School of American Research), J. Arnold (Cal Space), C. Keller (Los Alamos), and Dr. J. Johnson (Los Alamos Medical Center).

We would also like to thank the following people for their assistance in producing this book: Glenda Ponder, Mary Ann Olson, Luween Smith, Bessie Vigil, and Mary Ann Wright of the Los Alamos National Laboratory and Stan Holwitz, Shirley Warren and their associates at the University of California Press.

SECTION I

Resources: Human, Technological, and Cosmic

INTRODUCTION

Human destiny has been closely tied to technology for a very long time. If we were suddenly stripped of the fruits of technology, Earth could only support a few of us. Long ago in the tropics our naked ancestors did subsist off the bounty of nature without recourse to technology. Like gorilla and chimpanzee today, they lived literally hand to mouth. But some 5 million years ago we parted ways with our ape cousins and developed a life-style utterly dependent on technology. Even the most "primitive" of the surviving hunter-gatherer peoples are masters of technologies sophisticated by any standards in nature. Spears, digging sticks, bags and baskets, water containers, fire, clothing, shelters, and all the strategies and social organizations to use these form the technological base that makes this mode of human existence possible.

Thanks to the gifts of a series of technological revolutions—agriculture, industry, medicine, science—we are now more than four billion strong. The specialized food plants that give us such abundance could hardly survive without us. They are in a very real sense technological creations. Most of us live in cities and towns far removed from the basic resources of nature that support our very lives. Food, warmth, shelter—these basic needs are met by a web of technology that reaches from farms large and small, from oil fields and dammed rivers, from mines and factories into our homes. Human destiny is truly wedded to technology.

The technological prowess that has allowed the human tribe to spread over Earth and to grow so large now offers us the opportunity to leave our natal planet. The technology to leave Earth is at hand; the technology to live permanently in space is beginning to be developed. Of course, most of space, that grand residual category in which we Earth dwellers lump everything beyond Earth's thin atmosphere, is utterly hostile to warm-blooded, thin-skinned, and air-breathing creatures such as ourselves. To survive there we will need all our technological tricks. Yet we should not exaggerate the challenge. There are relatively hostile places here on Earth where human beings have long since learned, with the help of technology, to live satisfying lives.

We recall a down to Earth example from America's Bicentennial. July fourth was a short time away and the "Today Show" was about to do a broadcast from Wyoming, one of the last stops on NBC's fifty-state tour. A viewer wrote (we paraphrase), "Please remind any of your viewers who might be tempted by the beauty of our mountains to move here that Wyoming has two seasons: Winter and July." With those words in mind we might say that although living in space may demand some fairly high technology, so too does living the year round in the northern Rockies.

Although dreams of space travel are centuries old, at least as old it seems as the realization that there are other worlds than Earth, serious thinking about people living in space is a phenomenon of the twentieth century. As the century began, the founders of the Space Age, the Russian Konstantin Tsiolkovsky and the American Robert Goddard, sketched the first phases of space development. Goddard experimented with primitive rockets and dreamed of trips to the Moon and Mars. Tsiolkovsky, a nearly deaf schoolteacher, similarly felt the lure of the Red Planet. But he also described orbital colonies warmed by the abundance of solar energy that could be harvested in space. In thinking about the prospects of human settlements in space, rather than on the surface of planets or moons, Tsiolkovsky was ahead of his time. It would take the rest of us decades to catch up.

The successors of Tsiolkovsky and Goddard continued the process of learning how to make space travel and settlement a reality. Primitive rockets became V2s and then giant man-rated rockets that would carry a dozen human beings to the Moon. The work Goddard began has come a long way, but we have only made the first steps toward realizing Tsiolkovsky's vision of habitations in space. Skylab and the Salyut space stations are a beginning. Humanity is learning to live in space, albeit a few months at a time.

Specific designs for large human habitations in space appeared on the scene in 1974 when Gerard O'Neill published an article called "Colonies in Space." The beauty of O'Neill's concept was the combination of two ideas. One was the old vision of space stations. The other was a realization that there was a cheaper way to do it than hauling building materials up from Earth. Tsiolkovsky had long ago pointed out that energy is cheap in space. Abundant energy flows from the Sun and there is lots of room in space to collect it with very big mirrors, far more room than we have on Earth or could find on the other planets. O'Neill also reminded us that the other ingredient, raw materials, is more cheaply obtained from extraterrestrial sources than from Earth. The reason is that our planet is big and has a strong gravitational field.

One way to think about it is to consider just how much effort we expend fighting gravity in everything we do. We sit to give our legs a rest and sleep lying down so that our hearts need not work so hard pumping blood uphill to our heads. But we are rarely conscious of the price of living in a strong gravity field. We are accustomed to its influence. Gravity makes technology harder too. For example, the work of architects is more a matter of fighting gravity than anything else.

In space gravity plays a far smaller role. In orbit there are no appreciable effects of gravity at all. On the Moon gravity exerts only one-sixth the pull it does on Earth. On the Moon even puny human muscles can seemingly work wonders. The Apollo astronauts, though encumbered by bulky space suits and inexperience, hopped across the lunar surface with ease. When Alan Shepard, one of that very lucky dozen, hit the first lunar golf shot, he said (with some exaggeration) that the ball traveled "miles and miles." Other, even more spectacular feats await. Someday in the not too distant future a space traveler will do something science fiction writers have been talking about for decades: A human being will jump completely off a small world. Asteroids and small moons have very weak gravity indeed.

The strength of Earth's gravitational field makes the use of terrestrial materials to build structures in space an expensive proposition. If, as O'Neill suggests, we were to use lunar or asteroidal resources, the costs could come way down. Let us examine a useful illustration of the cost differential.

Earth-launched rockets are big for three reasons. One has to do with the design of chemical rockets, the need to carry fuel in the vehicle itself. Another is the friction a rocket experiences as it moves through Earth's atmosphere. However, the most important factor is the strength of Earth's gravitational field. Picture a Saturn Moon rocket with the command module and the lunar excursion module (LEM) perched at the very top. By far the smallest vehicle of the three is the LEM. Of all the tons of spacecraft assembled for a lunar mission, only a small fraction is the craft that takes two astronauts to the lunar surface and returns them to lunar orbit. The command module, only slightly larger, would boost its three man crew back to Earth. An enormous rocket was needed to lift the whole Apollo stack off Earth, but relatively tiny craft performed the lunar landing and return.

Imagine then that you need many tons of material in orbit to build a space station. There are two sources close at hand: Earth and the Moon. Without going into detail, the cost of getting a given amount of terrestrial materials to the building site would be about twenty times greater than the cost of transporting the same mass of lunar materials. Of course, there are other costs. If you only need a few shuttle loads in low Earth orbit, there isn't much point in going to the Moon (at least not at first). It would be more economical to fetch from Earth prefabricated feedstock for the machines that will make the structure of the space station. (Automatic construction robots are the way to go. It saves volume in the launch vehicle. Prototypes are already being developed by NASA contractors.) Another reason for using terrestrial materials for small near-term projects is that the industrial infrastructure is in place on Earth. You wouldn't have to pay enormous start-up costs to produce refined materials from raw ore. The industries already exist. On the Moon there is nothing but ore, a few dead spacecraft, and some "garbage" discarded by the astronauts.

But our need for building materials in space will not remain small for very long. Already we have orbited small space stations (Salyut and Skylab) and numerous communications and military satellites. As the need grows for larger facilities, especially bigger space stations, economies of scale will make the use of lunar materials more attractive.

Eventually (probably in only a few decades), we will look to the Moon for many of our needs. The key element of O'Neill's scheme for utilizing lunar resources is a device called a mass driver. Earlier we mentioned that one of the inefficiencies of rockets is the need to carry on-board fuel. That fuel provides the energy (and reaction mass) needed to reach orbit. If an external source could supply energy, the whole process would be simpler and far more efficient. The mass driver does just that (the enormous bulk of the Moon becomes the reaction mass).

A mass driver is a catapult or slingshot that uses electromagnetism as the driving force in much the same fashion as do electric generators and motors. A conductor carrying current in a magnetic field (a motor) or moving through a time-varying electric field (a generator) experiences a force. In O'Neill's mass driver a coil of wire embedded in the payload is pulled by electromagnetic fields through the coils of fixed accelerator stations. The mass driver could be powered by a rather modest array of solar panels and can accelerate packages of dirt to lunar escape speed. The solar array is the size of a few football fields, whereas the accelerator is only a few hundred meters long. The accelerations are large, too large for fragile payloads like people, but dirt does not really care what you do to it. (The lunar facility would, of course, have a small fleet of rather modest conventional rockets to ferry people to and from the lunar surface.)

O'Neill proposes launching ore (dirt) off the Moon rather than processed materials for the simple reason that it is easier to build processing facilities in orbit than on the lunar surface. Eventually, the economies of scale will permit building lunar processing facilities—out of lunar materials, of course. However, the economies of working in space rather than on the confining surfaces of planets and moons suggest that most human activities off Earth will be in orbit.

O'Neill envisions a modest lunar station to support the mass driver. A few kilograms of lunar soil could be launched every few seconds; the annual throughput would be thousands of tons. The big start-up costs would be transporting the mass driver and the solar array to the lunar surface. A few dozen shuttle flights would be sufficient. Once in place, the lunar facility would begin to deliver enormous quantities of raw materials to orbital processing facilities. There giant furnaces, also powered by solar energy, would reduce the lunar dirt to useful metals, to a slag invaluable as cosmic-ray shielding for people living and working in space, and to oxygen.

Oxygen is very useful stuff. Chemical reactions in which oxygen combines with other materials power our bodies and rockets. Because most of the mass of rocket fuel is oxygen, enormous savings in space operations can be realized by using lunar-derived oxygen to fuel the lunar ferries we mentioned earlier as well as orbit-transfer vehicles needed to haul the processed lunar soil from the smelters to the building sites.

With a mass driver-based lunar facility in place, large orbital facilities would become possible. Among these might be industrial plants that could eventually replace terrestrial heavy industry (why not take this major source of pollution off Earth entirely?) and orbital habitats, the real foundations of the human settlement of space.

Although the mass driver lies at the heart of O'Neill's concepts, it is mostly a means toward the human uses of space. With a ready source of building materials on the Moon, our presence in space could grow almost without bound. O'Neill has described orbital habitats, literal space cities that would each house thousands of people. A typical habitat might be kilometers across, with plenty of room for parklands and even rather fantastic landscaping on their internal surfaces. A breathable atmosphere would be confined within the habitat, whereas artificial gravity would be produced by spinning the structure.

If only one space city were ever built the cost would be fantastic, but seen in the context of larger space development the costs could be very affordable. A useful analogue is air travel. Every year millions of us travel great distances in airplanes at relatively low cost. But if every time a plane took off the passengers had to pay for new planes, runways, control towers, and the like, the airline industry would never have gotten off the ground. Fortunately, planes and airports have long lifetimes. Development and building costs can be spread over countless flights, bringing the price of tickets within common reach. Similarly, although there would be large capital costs to establish the space-based industries that could make large habitats a reality, those costs are comparable with the investments we have already made in the airline industry, in power plants, and in numerous other enterprises. Profits are waiting to be made in space; we only have to make the initial investment.

The end product of space development will be a large human presence in space, mostly in orbit. Like Europe's overseas colonies that started small and struggling, the space colonies must grow toward self-sufficiency.

One important shortcoming of the first stages of space development is the simple fact that the Moon, the nearest source of raw materials, is a large object. During its formation the Moon, like Earth, grew hot and melted. The heavy elements, metals like iron, nickel, and platinum, sank toward the center. Lunar soil like most terrestrial rocks is a poor iron ore. Also, because the Moon is much smaller than Earth, it has not retained an atmosphere, so the lighter elements, notably hydrogen and carbon, are also rare on the Moon. Fortunately for our long-term prospects in space, there are abundant extraterrestrial sources of these elements.

Four and a half billion years ago when the Solar System began to form, the most numerous objects were tiny bodies only a few kilometers across. Many coalesced to form the planets but others remained isolated and solitary. Today we recognize these remnants of Solar System formation as asteroids and comets. Being much smaller than Earth or Moon, they never melted and still retain their primordial burdens of heavy and light elements. Most asteroids occupy a belt between the orbits of Mars and Jupiter, whereas the vast majority of comets populate a realm far beyond the orbit of Pluto. Both populations constitute an enormous resource of materials for eventual human use. Because comets and asteroids have very weak gravity, the cost of extracting their materials is the cost of reaching them. No big rockets or even mass drivers are needed—only mining equipment. These are the bodies people could jump off powered only by human muscle.

Not all asteroids and comets orbit the Sun far from Earth. A few hundred are permanent residents of the inner Solar System and periodically pass close to Earth. One example, an asteroid called 1982 DB, will be in a very favorable position in the notable year 2001. A relatively minor energy expenditure (corresponding to a velocity change of about 100 m/sec or 225 mph) would bring us within reach of a potentially rich and economic source of extraterrestrial metals, carbon, and hydrogen. The latter would be in the form of ice.

Asteroids and comets may soon supplement and complement lunar resources for space development. Eventually, they may prove to be the stepping-stones for a human expansion throughout the Solar System: first, settlements in low Earth orbit supported from the surface of our planet; then orbital colonies built from lunar and asteroidal materials; a lunar colony; colonies planted among the near-Earth asteroids; a base on Mercury, the sun's most closely held planet, with a mass driver rather like its lunar cousin and used to launch building materials for gigantic power stations in close orbit around our star; a Martian settlement supported by earlier bases on the two Martian moons; colonies among the main belt asteroids; settlements on and among the icy moons of Jupiter, Saturn, and the other gas giant planets; and last, a multitude of colonies within the Sun's loosely held cloud of comets.

We will learn most of technologies we will need for this expansion into the Solar System during the next few decades. Space development will require human adaptation to a variety of new environments. During the initial phases it will be a difficult process, but we don't expect it to be different in kind from the sorts of technological and cultural adaptations we have experienced in the past. Like ourselves, descendants of those first hominids who left the sheltering African forests for a new life on the open savannas, our own descendants will be well equipped for the coming challenges in space.

In the first paper of this section, Finney and Jones review the career of the ultimate explorer, *Homo sapiens*. Whereas technology has played an important role in human development and will certainly play a central role in the support of human communities in space, we sometimes forget just how important our exploratory nature has been. Many of the things we do can be justified in economic, political, military, or religious terms. Visionaries of many ages, Columbus, Cook, Cheng Ho, Amundsen, von Braun, and countless others, may have worked their greatest miracles when they loosened official purse strings enough to get a project started. But the projects would never have been proposed without that key ingredient—the need to see what is on the other side of the hill—the need to try something for the simple reason that it has never been done before. It is an ancient trait, mankind's exploring urge; one as ancient and as intimately tied to the evolution of our species as upright posture, big brains, social behavior, and technology.

In the second paper, William Hartmann surveys the Solar System resources that will support space development. During the last decade, particularly since O'Neill's rediscovery of the value of lunar resources, many people's conceptions about space colonization have changed. Even as recently as the late 1960s, space exploration and space colonization meant planets and only planets. This is probably not surprising since we are all used to living on the surface of a big rock. However, as described earlier in this introduction, gravity means that planets are nice places to evolve but expensive places to live. The rich resources of materials and energy can be found in space, particularly on the small objects. Hartmann outlines current thinking about the formation of our Solar System, an intricate sequence of events that led not only to differentiation within a given planet (Earth's crust, mantle, and core being an example) but also to gross differentiation between the disparate provinces of the Sun's planetary system. For instance, among the small objects those richest in metals are closest to the Sun, whereas those farther from our star are composed primarily of ice. Such compositional differences may shape the course of human activities in space just as the sporadic concentration of rich ores on Earth plays a crucial role in global politics.

There are those who say that space development will never happen. Some say that it is too expensive; others that the global rivalry that seems to drive so much of space development will consume us all before we have really begun. Glen David Brin grants that we live in perilous times but, in the third paper, describes some small technological steps we might take now to make our foothold in space a little firmer. Brin describes what one might do in space with the Shuttle main tank, those enormous fuel tanks that we now discard in the Indian Ocean on every Shuttle launch. He summarizes arguments that show that with only a modest readjustment of mission operations, we could put the tanks of future missions into orbit. The payload delivered to orbit could actually increase. He then talks about some novel uses that could be made of the tanks.

From small beginnings large enterprises in space could grow. One significant factor that must be considered in any discussion of space development is that once a self-sufficient community is established, its potential growth is limited not by the resources of Earth or the carrying capacity of the terrestrial biosphere but by the vastly larger resources of the Solar System. One stunning trend evident in recent terrestrial settlement episodes (the colonization of North America and of Australia are excellent examples) has been the very high population growth rates sustained during the early years and decades. In the early 1800s the European population in Australia grew by 8 percent per year with similar rates implied for agriculture. Humanity is very capable of sustaining very high rates of development. In space where limits to growth may not be reached for centuries, the increasing use of sophisticated robots could, if we so choose, permit very, very rapid growth.

In the fourth paper of the section, David Criswell examines the potential for very large human undertakings in space. Criswell is well known for the breadth of his imagination, at least in the space science community. During the Los Alamos conference more than one social scientist commented that if Criswell achieved no other end, he made the prophecies of the other space scientists seem rather tame. Criswell has never shied away from the implications of his imagination, yet he remains firmly grounded in the limits that physical law places on the possible. In this regard Criswell is in such good company as Arthur C. Clarke and Freeman Dyson.

Criswell discusses the growth of human technology in terms of a concept called "cumulative controlled connectivities," which describes the interactions among matter, energy, and skill in the accumulation of human wealth and capabilities. It is a generalization that he applies equally to the formative phases of the Industrial Revolution, to contemporary human activities, and to space development. He also introduces the term s'homes (space homes) to describe the general class of orbital dwellings in which virtually all of our descendants may live. He ends his paper with a truly startling proposal. Sometime within the next few centuries or millennia, our descendants may have the capabilities and resources that would enable them to mine the Sun—not only that but also to reduce its size through a process called "star lifting" and thereby greatly increase its life expectancy.

Some of our nontechnical readers may find heavy going in Criswell's paper. Rather than sacrifice the rich detail of his contribution, we asked him to provide a summary of his chapter; that summary concludes his contribution.

No matter how the details of space development proceed in the Solar System, before too very long the lure of distant stars will draw the first voyagers into the interstellar deep. In the final paper of this section the editors return with a discussion of the technical means and the social implications of interstellar travel. Humanity's terrestrial migration used many means of transport. Most of the globe was settled by foot; some of it by horse or ocean vessel. Ships and airplanes and submersibles now take us to the last untouched corners of our globe. So too will our descendants spread out of the Solar System by several means. There is, we argue, one particularly intriguing alternative to reaching other star systems via fast spaceships: the slow road to the stars taken by colonizing wandering interstellar comets.

References

Burke, J. D., and A. S. Whitt, eds. *Space Manufacturing 1983.* *Advances in the Astronautical Sciences* 53. Univelt, San Diego.

Goddard, R. H., 1966. *Autobiography.* A. J. Ste. Onge, Worcester.

Goddard, R. H., 1970. *Papers.* McGraw-Hill, New York.

Kosmodemyansky, A. A. 1956. *Konstantin Tsiolkovsky: His Life and Works.* Foreign Languages Publishing House, Moscow.

O'Neill, G. K. 1974. The Colonization of Space, *Physics Today*, 27, 9 (Sept.). 32-41.

O'Neill, G. K. 1977. *The High Frontier: Human Colonies in Space.* William P. Morrow, New York.

O'Neill, G. K. 1981. *2081: A Hopeful View of the Human Future.* Simon and Schuster, New York.

Ordway, F. I., III, and M. R. Sharpe. 1982. *The Rocket Team.* MIT University Press, Cambridge.

"In space as on earth, technology must be cheap if it is to be more than a plaything of the rich."

—Freeman Dyson

"We can lick gravity, but sometimes the paperwork is overwhelming."

—Wernher von Braun

* 1

Ben R. Finney and Eric M. Jones

THE EXPLORING ANIMAL

We *Homo sapiens* are by nature wanderers, the inheritors of an exploring and colonizing bent that is deeply embedded in our evolutionary past. In this we are not unique; other species are also adapted for expansion. What makes us different from other expansionary species is our ability to adapt to new habitats through technology: We invent tools and devices that enable us to spread into areas for which we are not biologically adapted. As this technological capacity developed, it allowed our distant ancestors to spread over Earth and now enables us to contemplate leaving our natal planet. However, it is not simply the technological ability to build spaceships, life support systems, and the like that will drive the expansion into space. Whereas technology gives us the capacity to leave Earth, it is the explorer's bent, embedded deep in our biocultural nature, that is leading us to the stars.

To develop this thesis we have to start at the beginning of human evolution, at the appearance of the first tiny hominids in the grasslands of East Africa. We are hominids, a handy term for all erect-walking primates, a category that includes *Homo sapiens*, all extinct species of *Homo*, as well as all the species of more rudimentary two-legged creatures that preceded them. As Darwin pointed out, it is in Africa that our closest relatives, the chimpanzee and gorilla, live, and it is in East Africa that the oldest hominid fossils have been found. These date back some 3 million years but are not thought to represent the very oldest of hominids. Although some scholars have tried to push back the beginnings of Hominidae to some 10-15 million years ago, new techniques of comparing chromosomes, serum proteins, and hemoglobins between man and apes and of calculating the immunological distance between them indicate that the separation of the first hominids from our ape cousins may have taken place as recently as 5.5 million years ago.

The first "giant leap for mankind," to borrow Neil Armstrong's phrase, was the descent from the sheltering trees of the tropical forest to the open grassland-woodland environment of the savanna; a descent made by those as yet unknown ancestors who in so doing set the train of human evolution in motion. These were literally the first steps toward mankind, for they were made on two legs instead of four. This postural revolution left the forelimbs free to make and manipulate tools, to carry babies, food, and other goods, and to perform a myriad of tasks. It is this evolving capacity that made these first hominids and their descendants unique.

But this move into the grasslands was hardly, as some popular writers have imagined, an invasion of bloodthirsty hunters into the savanna. The earliest hominids known from the fossil record were small, generalized creatures, wholly lacking the ripping teeth or other natural adaptations of successful predators. For example, the oldest nearly complete skeleton known, discovered in Ethiopia, is that of the famous Lucy who stood a bare three and a half feet tall, weighed a scant 60 pounds or so, and had a set of almost humanlike teeth. Without a highly sophisticated hunting technology, such a modest creature could hardly have topped the savanna food chain; indeed, the archaeological record indicates that such a technology did not develop until several millions of years after the move from the forest into the savanna.

How then did these tiny hominids survive and prosper?

They became the premier food gatherers of the savanna. Their bipedal posture, with that crucial freeing of the hands, enabled them to tap a wide range of grassland resources: They probably gathered nuts, berries, birds' eggs and grubs; dug up succulent tubers and roots; caught insects and small animals; and perhaps also ambushed the young of larger animals. Yet they did not accomplish all this with their bare hands. In fact, these early hominids must have been the first creatures dependent upon technology, however rudimentary, for their survival. Unfortunately, the hard archaeological evidence has not survived. The most crucial tools, probably made of wood, fiber, or skin, were digging sticks, simple containers, and other rudimentary implements to aid the gathering of food. A new economy was now possible. With these simple tools, mature, able-bodied males and females could range over the countryside in search of food. Then instead of consuming it on the spot, they could, thanks to their erect posture and free hands, carry the food back to a base camp to share with dependent children and adults who had stayed behind. This new food-gathering way of life thus led to a home-based social organization with all its implications for family .formation, prolonged nurturance and training of the young, and sharing and communication.

For all their evolutionary advances, these early hominids, generally classified as members of the genus *Australopithecus*, apparently did not expand beyond the savannas of Africa. To migrate farther required, it seems, further evolution. Within the *Australopithecus* genus there was speciation. Arguably, the first known hominid species, *Australopithecus afarensis* (to which Lucy belongs) was followed by at least two successor species that survived until 2 to 1.5 million years ago: a gracile type that developed further the generalized omnivore niche pioneered by its distinguished ancestor and a robust type which, as witness its massive jaws and molars, must have specialized in a diet of coarse, gritty tubers and roots. Although a few paleoanthropologists, notably those from the famous Leakey family, reject direct descent from any known *Australopithecus* species, most see further evolutionary advance in Hominidae as coming from either *afarensis* or its similarly gracile descendant.

Paleoanthropologists speak of mosaic evolution, of the accelerated evolution of parts of the body while other features remain relatively static. Thus, although *Australopithecus* made the tremendous advance to erect posture, with all the modifications of the feet, legs, and pelvis that required, over the 2-3 million years the genus is known, to have existed, its brain remained small, averaging around 500 cc, hardly bigger than that of its chimpanzee cousins. Then starting about 2 million years ago, evolution of the brain began to accelerate. The first evidence of this trend comes from the skull of so-called Handy Man (*Homo habilis*) discovered at Olduvai Gorge, Tanzania, by the late Louis Leakey. Although some students would classify it as an advanced *Australopithecus*, most have accepted Leakey's assignment of it as the first known representative of our genus, *Homo*. This is because of both its significantly greater brain capacity of 650 cc and its undeniable association with worked stone tools. Although recent discoveries might seem to confirm the long-held conjecture that *Australopithecus* must have used rudimentary stone tools, by the time *Homo habilis* appeared the distinctly human synergy between the development of increasingly sophisticated tools and the acceleration of brain development seems definitely to have been underway.

Although to some scholars skull fragments unearthed on the island of Java indicate that *Homo habilis* may have been the first hominid to leave Africa, the next species of *Homo* to evolve, *Homo erectus*, is generally credited with having been the first hominid to spread in any numbers beyond Africa; its fossil remains have been found widely scattered over Eurasia. In fact, the first *erectus* fossils were found not in Africa but far away in what is now Indonesia (Java Man), China (Peking Man), and Germany (Heidelberg Man).

Homo erectus was significantly brainier than his predecessors. Fossil *erectus* skulls range in brain capacity from around 775 cc to 1,225 cc, overlapping the low end of the *Homo sapiens* range. This advanced hominid employed a more highly developed stone technology—inventing, apparently, the art of chipping stone on both sides to make a keener edge—and was a successful big game hunter. This involved a critical shift in the savanna niche pioneered by the *Australopithecus* ancestor—from that of a food gatherer who also caught some lizards, birds, and other small animals to that of a hunter and gatherer who, in addition to harvesting wild vegetable foods, began to prey systematically on large herbivores. This shift may have had an important physiological dimension. If our relative hairlessness and abundance of sweat glands, and hence our outstanding ability to dissipate heat through copious sweating, evolved at this time, *Homo erectus* hunters could, unlike other predators that hunted in the cool of the late afternoon or evening, have operated in the heat of the day, catching prey unawares or running them to exhaustion. But, above all, the shift to a hunting emphasis had a specifically cultural dimension. Hunting technology, involving both tools and organization, now comes to the fore. For example, as can be reconstructed from excavated kill sites, these hunters skillfully employed guile and teamwork to drive large animals, even elephants, into bogs or other traps where they could be slaughtered with spear or club and then butchered with finely chipped cutting tools.

This hunting adaptation enabled small bands of *Homo erectus* to wander north, over many generations, out of Africa and then pursue game east and west over the warm savannas, which at the beginning of the Pleistocene some 1.5 million years ago stretched the length of

South Asia and into southern Europe. Once in Europe and Asia, however, these hairless, tropically-adapted hominids would have been subject to periodic cold stress as the glaciers began to form and periodically advance southward. Yet archaeological evidence indicates that *Homo erectus* bands roamed far north in treeless grasslands rich in game but so much colder than the African savannas. Fire-blackened hearths, dating as far back as 700,000 years, found at some of these sites reveal that these hunters had learned to control fire—one of the single most important innovations in cultural evolution. With the ability to keep warming fires burning and to gain further protection from rude shelters and animal skins, *Homo erectus* was able to penetrate far north, reaching at least latitude 49° north during interglacial periods.

For all his hunting skill and cultural ingenuity during the million or so years of his existence, *Homo erectus* did not apparently succeed in spreading beyond the linked continents of Africa, Asia, and Europe. All the available evidence indicates that the move to the Americas and Australia followed further cultural development and the evolution of a new species, *Homo sapiens.*

Details of the origin of *Homo sapiens* are far from clear. Fossil skulls found in western Europe that date back some 250,000 years show an unmistakable trend toward greater cranial capacity and the high, vaulting shape of modern skulls—with all that implies for increased mental capacity. Yet in Europe at least, the gap between these evolving skulls and the late appearance around 40,000 years ago of modern *Homo sapiens* is filled with the abundant remains of the famed Neanderthal Man whose projecting face, beetle-brows, and thickset build once seemed to belie any smooth progression to modern forms. In fact, until recently many scholars classified Neanderthal as a separate species, an evolutionary dead end. Now, however, because of better reconstructions of his skeletal remains, a realization of the fact that at 1,600 cc his brain was slightly larger than the average for a modern man, and an appreciation of the possibilities for microevolution under climatic stress, many students are inclined to classify Neanderthal as an early form, or subspecies, of *Homo sapiens* physiologically adapted to the bitterly cold conditions of the late Pleistocene. But even this rehabilitation of Neanderthal does not solve the mystery of exactly where and when *Homo sapiens* originated. Was it somewhere in Europe or Asia, or was Africa the cradle of modern man as well as the ancestral forms?

Whatever the case, for our purposes the important point is that *Homo sapiens* apparently were the first to populate the hitherto empty continents. The drastic lowering of sea levels by 80 to 100 m during the last glaciations of the Pleistocene facilitated this movement—by exposing the continental shelves so that Siberia and Alaska were joined by a land bridge, while Indonesia was made into an extension of Asia reaching out almost to the shores of a greater Australia composed of the present continent, New Guinea, and surrounding shelves. Yet previous glaciations had similarly lowered sea levels without any hominid migrations taking place. The crucial ingredient was the evolution of human cultural capacities and techniques. Refined hunting tools, tailored skin clothing, and other survival gear enabled *Homo sapiens* to penetrate the Arctic; then all that had to be done to reach America was to follow prey across Beringia (as geologists dub the broad plain that then linked Asia and North America). Similarly, once simple rafts and rudimentary techniques for living off sea and coastal resources had been developed, people could cross the narrow stretches of open water then separating Sundaland (glacially enlarged Indonesia) and Sahuland (greater Australia).

Exactly when this took place is still subject to debate. Previously, scholars thought that these movements could not possibly have taken place earlier than 12,000 years ago and probably more recently. Now, however, the discovery of respectably ancient human fossils in the middle of Australia leads many archaeologists to estimate that people first crossed to Sahuland during the last glaciation some 50,000 years ago. Some even suggest that the first crossings might have been accomplished as early as 120,000 years ago. Although the status of supposedly ancient remains in North America is still subject to dispute, similarly early (and even earlier) estimates of the first crossings of Beringia are being increasingly voiced.

Whatever the exact dates, in surmounting tropical and arctic barriers and then in spreading over the forests, mountains, plains, deserts, and jungles of the three new continents, these ancient wanderers utilized the unique human ability to adapt culturally to new environments. Building on the biological foundation of erect posture and brain expansion, our more recent ancestors added the capacity to invent and apply technology to make human existence possible from Africa to the Americas, from the tropics to the arctic. Where other animals had to evolve biologically to move into habitats radically different from the ones for which they were specifically adapted, *Homo sapiens*, the hairless biped from the African savanna, could adapt culturally.

Then as the glaciers receded, ocean levels rose, and expanding populations pressed against resources, people were forced to intensify their food quest by learning how to grow crops instead of simply gathering them and how to raise animals instead of just hunting them. Among the arts that flourished with the advent of this agricultural revolution was seafaring. As coastal people developed more seaworthy vessels and more trustworthy methods of navigation, they began to venture farther and farther out to sea.

By A.D. 1000 the Polynesians and their Micronesian cousins, probably the first people to sail into the heart of any ocean, had discovered and settled just about every inhabitable island within the vast Pacific. About the time they were completing their migrations, first Irish then Norse seafarers spread over the islands of the North Atlantic. But the true discoverers of the global sea were those European navigators who, a few centuries later, learned how to sail regularly between continents and eventually around the world. This recent European expansion, however hard it was on the peoples of the world who bore the brunt of the *conquistadores*, traders and the diseases they brought, did serve to bring together the long separated branches of humanity into one world system. After some five million years of hominid biological and cultural evolution, *Homo sapiens*, the sole surviving hominid species, had not only spread all over the world but also was on the way to forming a single, intercommunicating population with genes, ideas, and artifacts flowing back and forth across oceans and continents.

Today, because of continued economic and technological growth, we stand on the threshold of space. Although it may be easy for some to dismiss the dreams and designs of colonizing space as mere extensions of Western imperialism or of technological thinking gone wild, we maintain that the urge to expand into space is basic to our human character. We are the exploring animal who, having spread over our natal planet, now seeks to settle other worlds.

Of course, in a general sense virtually all vertebrates, and many invertebrates as well, must also be considered as exploring species. To survive, animals must explore their environment to find sources of food and living space, and a successful species is one that expands its habitat through migration. What makes us different is that in hominid evolution this basic urge to explore has been developed further, to the point where it is leading us to leave Earth.

Man is an animal that has professionalized exploration. It is the juvenile of most animal species who do the exploring, the investigating of their environment before settling down on a limited geographical range from which as adults they hardly stir. Modern man, from Australian aborigines to the denizens of an industrial city, follow a similar pattern of juvenile exploration—of the waterholes and

sacred places of the desert or of the sights and experiences of touring
Europe or backpacking in the Sierras—before settling down to the
routine of adult life. Yet some adults do not give up their exploratory
bent and in fact make a career of it. Columbus did this through sheer
entrepreneurial genius; by the late eighteenth century maritime exploration had matured to the point that Captain Cook could claim to
be "employed as a discoverer." Now we even have people who make
their living by exploring the stars and planets through telescopes and
robot spaceships and a growing corps of astronauts, cosmonauts, and
spacionauts who actually explore space in person. We are the animal
that has turned a juvenile characteristic into an adult passion.

This is as much part of our genetic evolution as it is of our cultural
progress. Our hypertrophied exploratory urge stands out as a
behavioral manifestation of our curiously retarded development.
Anatomists have long remarked how adult humans resemble juvenile
apes in their large brains, globular heads, and lack of protruding
muzzle. Some have proposed that we have become brainy humans
through mysterious changes in growth rates that act to preserve, far
into human maturity, fetal or juvenile characteristics. At birth infant
apes and humans are much alike in brain size and facial configuration.
But whereas the sutures close early in an ape's skull and its brain
grows little more as its brow thickens and its jaw and teeth develop
into a protruding muzzle, our sutures remain open and our brain
continues to grow through a prolonged infancy and adolescence. We
reach maturity with underdeveloped jaws and teeth tucked in-
conspicuously under the bulbous brow of childhood, a trick of nature
that has in a relatively short span nearly trebled the size of our brain.

Ethologist Konrad Lorenz boldly extends this theory of human
evolution through neoteny into the behavioral realm. He notes how
humans retain a range of juvenile behavioral traits into adulthood,
notably the penchant for investigating and exploring their environment. Unlike our ape cousins, especially the dour gorilla, we—or at
least some of us—retain our childhood curiosity into maturity. This
retardation, according to Lorenz, has served the species well; from it
flows our inquisitiveness into the nature of things as well as our
incessant searching for what lies over the horizon—for, in other
words, science and exploration. Through this process of evolution by
retardation, now made less mysterious by the realization of how the
mutation of just one or a few regulatory genes could affect such a
radical restructuring, we have become a most inquisitive, exploring
animal.

In fact, we hypothesize that our exploratory bent will be as crucial
to our future evolution as it was to our past development. If the

technology of space colonization really works, if our descendants do settle throughout the Solar System and then migrate to other star systems, humanity will never be the same again. The course of human evolution will change utterly and inalterably. This is because, by scattering through the vastness of space, our descendants will be setting up the conditions necessary for the rapid speciation of *Homo sapiens*. The threshold of space is also the threshold to quantum biological evolution.

Stephen Jay Gould and other paleontologists maintain that major evolutionary divergence proceeds through bursts of speciation, through the comparatively rapid splitting off of separate lineages from the ancestral stock, and not by the gradual transformation of that stock. According to this increasingly accepted view, rapid speciation occurs primarily in very small populations that have become geographically isolated from the ancestral stock. Where genetic change is resisted by large populations well adapted to their environment, favorable genetic mutations can easily gain a foothold in marginal geographic areas where pressures for natural selection can be intense and then spread quickly through the small populations that have become isolated there.

The pace of hominid evolution was undoubtedly forced by the spread of bands of hunters and gatherers first through Africa and then across Eurasia. Radical climatic shifts, tectonic events, and the tyranny of distance then acted to isolate small subpopulations in out-of-the-way corners of these continents, providing the opportunity for adaptive mutations to take hold. Then as selection proceeded, new species emerged. But this once vigorous speciation response that brought forth so many species of *Australopithecus* and then *Homo* has now been greatly dampened. Now because we, the surviving hominid species, form one intercommunicating, interbreeding world population, rapid speciation would no longer seem possible.

But that holds only if we stay on Earth. If our descendants spread far and wide through space, the forces of evolution now braked on Earth will be released once more. As they scatter through the Solar System and eventually across the gulf of light years to other star systems, our descendants will experience that prerequisite for rapid evolution our ancestors once knew: isolation in small and distantly separated communities.

Advances in genetic engineering may further accelerate the pace of evolution, but again only in space. On Earth and in nearby colonies it is likely that any radical restructuring of the human form or psyche will be greatly restricted if not prohibited outright. However, isolated communities many light-years away will be freer to experiment in creating beings more adapted to the new environments, for they will be far from the blandishments of the crowd from which they had fled and beyond the range of any genetic police.

Human evolution in space will hardly be limited to the birth of one new species. Space is not a single environment but an Earthcentric residual category for everything outside our atmosphere. There are innumerable environments out there providing countless niches to exploit, first by humans and then by the multitudinous descendant species. By expanding through space we will be embarking on an adventure that will spread an explosive speciation of intelligent life as far as technology or limits placed by any competing life forms originating elsewhere will allow. Could the radiation of evolving, intelligent life through space be the galactic destiny of this Earth creature we have called the exploring animal?

References

Baker, Robin. 1980. *The Mystery of Migration*. Macdonald, London.

Boaz, Noel T., F. C. Howell, and M. L. McCrussin. 1982. Faunal Age of the Usnu, Shungura B and Hadar Formation, Ethiopia. *Nature* 300, 5893: 633-635.

Brace, C. L. and A. Montagu. 1977. *Human Evolution*. Macmillan, New York.

Campbell, B. G. 1982. *Humankind Emerging*. Little, Brown, Boston.

Darwin, Charles. 1871. *The Descent of Man and Selection in Relation to Sex*. John Murray, London.

Dobzhansky, T. 1962. *Mankind Evolving*. Yale University Press, New Haven.

Eisely, Loren. 1957. *The Immense Journey*. Franklin Watts, New York.

Eldridge, Niles, and S. J. Gould. 1972. Punctuated Equilibria: An Alternative to Phyletic Gradualism. In T. J. M. Schopf, ed., *Models of Paleobiology*. Freeman, Cooper, San Francisco. Pp. 82-115.

Goldschmidt, R. 1940. *The Material Basis for Evolution*. Yale University Press, New Haven.

Gould, Stephan J. 1977. *Ontogeny and Phylogeny*. Harvard University Press, Cambridge.

Isaac, Glyn L. 1976. East Africa as a Source of Fossil Evidence for Human Evolution. In G. L. Isaac and E. R. McGown, eds., *Human Origins*. Benjamin/Cummings, Menlo Park, California.

Johanson, Don, and M. Edey. 1981. *Lucy: The Beginnings of Humankind.* Simon and Schuster, New York.

Johanson, Don, and T. D. White. 1979. A Systematic Assessment of Early African Hominids. *Science* 205, 4378: 321-324.

Leakey, Louis S. B., P. V. Tobias, and J. R. Napier. 1964. A New Species of the Genus Homo from Oldavai Gorge. *Nature* 202: 4927: 7-9.

Lorenz, Konrad. 1971. *Studies in Animal and Human Behavior*, vol. 2. Harvard University Press, Cambridge.

Lowenstein, J. M. 1982. Twelve Wise Men at the Vatican. *Nature* 299, 882: 395.

Sarich, Vincent M., and A. C. Wilson. 1967. An Immunological Time-Scale for Hominid Evolution. *Science* 158, 3805: 1200-1203.

Stanley, S. M. 1979. *Macroevolution: Pattern and Process.* W. H. Freeman, San Francisco.

Tanner, Nancy M. 1981. *On Becoming Human.* Cambridge University Press, Cambridge.

✳ 2

William K. Hartmann

THE RESOURCE BASE IN OUR SOLAR SYSTEM

Just as processes of terrestrial crustal evolution have produced concentrations of useful materials at different parts of Earth's surface, general processes of planetary evolution have produced concentrations of different resources in different parts of the Solar System. The process of planetary formation produced a zonal structure, ranging from metal-rich silicates near the Sun through concentrations of organic and rocky material in the mid-Solar System to concentrations of ices in the outer Solar System. Melting has also concentrated metals in asteroidal cores, later exposed by collisions and fragmentation of asteroids. As a result of gravitational perturbations, samples of all these materials pass through the inner Solar System where they will be used for human activities in space. Planning is already underway to see how we can use these materials for the benefit of terrestrial and interplanetary civilization and to benefit Earth's environment at the same time.

The first steps in human migration off Earth are being taken during our generation. They include not only the first footsteps on the Moon, which have already happened, but also the first use of resources in space, which is now being planned.

In utilizing Earth resources in recent centuries, we have swept around our globe, looking for the most accessible deposits of ores and fuels. When these were used, we had to dig deeper and deeper, accepting lower grade materials that were more and more costly to process in both dollars and resultant pollution. Historically then, resources have become more expensive in final product cost in the marketplace and in damage to Earth.

At the same time, costs to operate in space have decreased because we are developing more capabilities for routine space operations. These are being paid for out of different pockets: military budgets, scientific exploration programs, and government programs to enhance national prestige.

In future decades there will be a crossing of these two curves: the rising cost of material acquisition on Earth and the falling cost of acquiring equivalent or substitute materials in space. This will provide the economic incentive for large-scale acquisition and utilization of space resources. This article addresses what materials are available, with emphasis on asteroidal resources, and how we can exploit them to the best interests of humanity.

Cosmic Differentiation: The Resource Pattern in the Solar System

In using Earth resources we had to find the environmental niches in Earth's crust where hydrothermal action, melting, sedimentation, burial of ancient forests, or other processes concentrated the materials we wanted. On each individual planetary body there may be a similar search for special locations, such as outcrops of nickel-iron core material on the surfaces of collisionally broken asteroids. The process of concentrating certain materials is called *differentiation* by geologists. Mining geologists deal with processes of localized differentiation in Earth's crust.

A much vaster pattern of cosmic differentiation occurs across the Solar System as a whole for which there is no equivalent on Earth or other individual planets. The process of Solar System formation produced compositional zones at different distances from the Sun. The inner five worlds—moonlike Mercury, hot, cloudy Venus, benign Earth with its moon, and cool, dusty Mars—are rocky bodies with various proportions of metal. Mercury, closest to the sun, has the most metal and the highest density. Between Mars and Jupiter lies the asteroid belt, which marks a transition zone. The asteroids of the *inner* belt are also rocky, and spectroscopic studies with Earth-based telescopes show that there are various subtypes of different compositions. Some apparently underwent intensive heating and melting. These produced surfaces of volcanic, lavalike rock and interiors of metal. Others have rocks that apparently never melted, preserving traits of primitive Solar System materials. Collisions broke open many of these varieties of metals and varieties of interior rocks. As we go toward the outer belt we encounter a new type of asteroid material composed of carbon-based minerals. Many of these minerals contain water, not in liquid form, but chemically bound within the minerals.

Beyond the asteroid belt lies the realm of giant planets. In this realm, beyond a point between the asteroid belt and Jupiter's orbit (that is, beyond about 4 AU (astronomical units) from the sun)* ice is relatively stable, even if exposed to direct sunlight. Therefore, ices are common in this part of the Solar System. Dominating this region is Jupiter, the largest planet, roughly ten times the size of Earth. Jupiter has a faint ring and a family of at least sixteen moons that are typically composed of roughly 50 percent rock and 50 percent frozen water. Next is Saturn with its beautiful ring system of ice particles and its large family of at least seventeen moons, which are mostly composed of water ice. Beyond Saturn is Uranus, also with a ring system and five known moons, and then Neptune with two known moons. Ices composed of frozen methane and ammonia dominate these outer bodies.

As we will see in later sections, the moons of the giant planets offer a host of unique worlds, with features ranging from active volcanoes with 100-mile-high eruption clouds, featureless ice plains floating on a sea of water, cracked and grooved ice plains, a cloudy atmosphere overlying a possible ocean of liquid ethane and methane, a possible sea of liquid nitrogen, and a moon with one black and one white side. In spite of these dramatic differences, the overall trend in composition as we go out from the sun is from stone worlds to ice worlds. In more detail, it is from metal-rich stone through metal-poor stone, carbonaceous stone, ice-stone mixtures, and familiar ice (frozen water) to unfamiliar ices of frozen methane and ammonia.

This pattern was caused by the process of planet formation in a disk-shaped, cooling gas cloud that surrounded the Sun soon after it formed. This cloud is called the solar nebula. Consider a cloud in our own sky, cooling as it rises in the atmosphere. The condensable

---

*An astronomical unit (AU) is a measure of distance; it is the average distance between Earth and Sun—about 93 million miles or 150 million kilometers.

material in the cloud—water—condenses to form water droplets, solid particles like snowflakes, or hailstones. In the same way, condensable substances formed solid grains in the solar nebula as it cooled. The important point is that the solar nebula was hotter near the sun and colder in its outer regions, and the local temperature controlled what could condense. In the hottest regions only high-temperature minerals, such as metal-rich silicates, could condense. A little farther from the sun, moist carbonaceous (carbon-rich) minerals could condense, and these were added to the mixture of rocky silicates. In the outer Solar System it was so cold that ices were added, and the ices were so abundant that they swamped the stony material. Eventually, the gas of the nebula blew away, leaving the solid grains behind to aggregate into planets. The planets of the inner Solar System were thus formed of familiar rock. We see in the asteroid belt a transition to carbonaceous rocks with water of hydration. In the Jupiter zone we have half rock and half water ice. We find mostly water ice in the Saturn zone. Among Uranus and Neptune we find ices of methane and ammonia playing important roles.

Thus, the Solar System is differentiated on a large scale, independent of any local variations on individual planets. The compositional zones guarantee supplies of a variety of materials from metals to frozen water. Humans, who started venturing into space for knowledge and national prestige, will soon venture into space to acquire materials, partly to support even more space exploration. This feedback loop will foster more migration into space: Early materials will support colonies that will be able to process other materials. Of special interest as material sources will be the Solar System's smallest bodies: asteroids and comets. Their negligible gravity will make it easier to haul materials from them to space stations than to launch the same materials "uphill" from the high-gravity surface of Earth (or other planets) to the same space stations.

Asteroids and Comets: A Variety of Accessible Materials

To understand the variety of asteroidal and comet materials available in near-Earth space, we must first review the emerging "big picture" of spatial distribution among asteroidal and cometary compositional classes scattered across the Solar System. By the early 1970s spectrophotometric analysis of asteroid spectra revealed compositional differences from one asteroid to another. The crude compositional data could be obtained from spectral absorption bands, especially the near infrared absorptions of important meteoritic and igneous minerals such as olivine and pyroxene. Differences also appeared in the general colors, some asteroids being rather neutral in color and others being quite reddish. Measurements of the thermal emission (light and infrared radiation) from asteroids soon allowed observers to calculate the albedos or reflectivities of specific asteroids. The combination of spectral bands, colors, and reflectivities showed the variety of asteroids in ever-increasing detail. Observers soon developed a somewhat arbitrary taxonomy of letter designations for the different asteroid spectral classes. Meanwhile, laboratory researchers compared the spectra of asteroids with the spectra of meteorite samples and found similarities between certain types of meteorites and certain spectral classes of asteroids.

For example, the C-type asteroids were found to have relatively flat spectra and very low reflectivities of only around 5 percent, and their spectra were found to resemble those of black, carbonaceous chondritic meteorites. Other relationships are less clear. The common S-type asteroid was believed to be stony or stony-iron material, possibly resembling the most common meteorite types, the chondrites or certain stony-irons. M-type asteroids were suspected by some researchers to resemble stony-iron meteorite materials, although the spectra of the metallic phase is relatively featureless and therefore hard to confirm. Cruikshank and Hartmann recently found that the rare A-type asteroids (five examples known) have spectra showing only olivine with no pyroxene, and they may be examples of stony-iron objects composed of olivine and NiFe (Nickel-Iron) metal.

Even within the asteroid belt, the compositional zones of the solar system can be traced by asteroidal spectral classes. E-type asteroids (possibly enstatite-chondrite materials) are concentrated near the inner edge of the belt, and the S types and M types are prominent in the central part of the belt. The C-type asteroids, which resemble carbonaceous chondrites with their lower temperature, volatile materials, and abundant water of hydration, are located in the outer belt and in the two clouds of Trojan asteroids occupying the Lagrangian points 60° ahead of and behind Jupiter, in that planet's orbit. Also prominent among the Trojans is a newly identified D-type asteroid that has very low albedo (probably indicating the presence of the low-temperature carbonaceous materials) but a much redder color than the C-type asteroids. (The D classification comes from the origin acronym RD, which stood for reddish-dark, describing the spectral characteristics. Asteroidal classes were subsequently reduced to single-letter names.) The reddish color has been attributed by Gradie and Veverka to organic compounds formed at even lower temperatures than the suite of normal C-type, carbonaceous materials. Several small outer moons of Jupiter have been identified as also exhibiting the dark, carbonaceous spectral characteristics and may resemble the outer Solar System asteroids.

As we go farther out into the Solar System, ices become increasingly important among surface materials of small bodies. We know that comets come from a swarm of objects that originated somewhere in the outer Solar System; the defining cometary properties of gas emission and eruptive instability are, of course, a result of their ices. Thus, we can view the entire population of asteroids and comets as grading from rocky and metallic-rocky materials in the inner Solar System and belt through rocky objects dominated by carbonaceous materials and organic compounds, possibly mixed with ices, to comet nuclei in the outer Solar System, which are dominated by highly reflective ices. Lebofsky has not only identified the water of hydration of some C-type asteroids but has also raised the possibility of a detection of frost on the surface of Ceres, the largest C-type asteroid.

Both theoretical and observational studies have suggested the likelihood that at least some asteroids contain surfaces of regolith material; that is, finely granulated powders produced by meteorite impact rather than pure, coherent rock surfaces. This result is important in contemplating sampling and refining of resources; a deep regolith would obviate the need to cut apart or crush chunks of coherent asteroid. The depth of the regolith is highly uncertain.

In summary, asteroid science has revealed a wide range of potentially interesting materials in space: rocky material, metals, organic materials, and ices, existing either in powdered or solid form.

The Near-Earth Sample of Asteroids

The asteroid belt and the different suites of comets and asteroids in the outer Solar System are reservoirs from which our samples of near-Earth bodies are drawn, partly at random. The drawing is done by processes that we do not fully understand. These include:

1. Jupiter resonances,* which throw objects out of the mid-belt;

---

*A gravitational resonance occurs at places in the asteroid belt where the ratio of asteroidal orbital periods to Jupiter's period is the ratio of small integers (for example, 2/3, 3/4) so that the asteroid is frequently closest to Jupiter at the same phase in their respective orbits. Such a situation produces strong orbital perturbations. An asteroid that wanders into a resonant orbit is soon thrown out of it. The gaps in Saturn's rings are produced in the same way; the gaps are resonant with respect to Saturn's inner moons.

2. Mars gravitational perturbations, which throw objects out of the extreme inner edge of the belt;

3. Giant planet perturbations, which send outer solar system debris into the inner Solar System.

Thus, the asteroids that cross Earth's orbit are a heterogeneous mixture of different types of objects from different source regions.

Also important in our sampling is the size distribution of objects in the asteroid and comet reservoirs. Generally, the number increases dramatically as the size decreases, and there are many more small objects than large ones. Typically, if the size is decreased by a factor of 10, the number of objects increases by about a factor of 100. So the sample of near-Earth objects includes dozens of relatively small asteroids (mostly with diameters from 0.5 to 10 km) of mixed compositions. Most probably originated in the asteroid belt, but a substantial fraction may be burned-out comets. Most researchers believe that the latter could include carbonaceous stony objects with some volatiles. The sequences of perturbation processes can put both asteroids and burned-out comets onto Earth-crossing orbits.

The objects that cross Earth's orbit are called Apollo asteroids named for the first such object discovered. The perturbation processes that throw asteroids and comets onto Apollolike orbits create objects that can approach Earth at relatively low velocity; hence, these objects are easy to reach from Earth because a rocket needs to make only a minimal velocity change (so-called ΔV)* to go from a given starting orbit near Earth to the surface of the asteroid. Indeed, it is easier to reach at least four of the ten closest asteroids (as taken from a 1983 list) and return with material from their surfaces than it is to reach the Moon and return with material from its surface. This is because the asteroids have negligible gravity and require little fuel for landing or lift-off.

In addition to Apollo asteroids, active icy comets and a few "burntout" comets, depleted in ice, cross Earth's orbit. The active comets generally have higher velocities and are harder to reach than Apollo asteroids. However, a few comets and a number of hypothetical burntout comets may approach Earth at low velocity and would be prime candidates for acquiring ices—water supplies—in nearby space.

---

*ΔV is the velocity change needed to rendezvous with an asteroid or another orbiting spacecraft. ΔV is a measure of the energy cost of a space mission. Asteroid missions are attractive because many have a total ΔV less than a lunar landing and return.

Thus, in the early stages of human migration into space, we can predict a vigorous interplanetary economy in which materials are extracted and shipped back and forth among asteroids, planet-orbiting colonies, and bases on various satellites and planets.

Anticipated Apollo Discoveries: More Resources in the Inner Solar System

An important principle derives from the preceding facts. The near-Earth asteroid inventory (June 1983 listing by TRIAD—Tuscon Revised Index of Asteroid Data) currently includes 27 Apollos (that come inside 1 AU) and 39 objects that come inside 1.1 AU. Many are larger than ~1 km radius. The current discovery rate is several objects per year, thanks primarily to survey programs. We can confidently expect a growing sample of kilometer-scale potential target asteroids in the near future.

A more subtle point, less widely appreciated, is that the diameter distribution virtually guarantees that there must be thousands of smaller objects that have not been detected but that would make interesting mission targets. For example, for every thirty 1 km Apollo asteroids, we might anticipate some three thousand 100 m objects. This increased sample would undoubtedly include some targets of much lower ΔV than those now known. As our search sensitivity increases, we can confidently expect a rapid increase in cataloged small bodies that could be used for resources or even be maneuvered into Earth orbit as resource reservoirs.

Composition of Earth-Approaching Objects

From the meteorites that reach the ground, we know something about the interplanetary population of objects near Earth. As shown in table 2.1, 81 percent of these are relatively primitive chondrites, 9 percent are achondrites (igneous rock that have undergone melting), 5 percent are carbonaceous chondrites, 4 percent are irons, and 1 percent are stony-irons. A problem with these statistics of ground-collected objects is that carbonaceous chondrites are very weak and crumbly, and a large fraction of them may not survive passage through the atmosphere. Thus, carbonaceous chondrites may be under-represented in our ground-based sample relative to their interplanetary abundance. Indeed, the fraction of carbonaceous objects in interplanetary space may be much larger than the 5 percent fraction reaching the ground. Consistent with this is the observation that the meteoritic component in the lunar soil is carbonaceous and the observation from meteor cameras that many small meteors are weak and burn up in the atmosphere.

The most thorough inventory of spectral data on compositions of near-Earth asteroids is that of McFadden. From a list of seventeen objects with good spectral data, she was able to make only six identifications with meteorite types, as shown in table 2.1. Of these,

TABLE 2.1   Best Examples of Meteorite Analogs Among Near-Earth Asteroids

Meteorite Type (and % of Observed Falls)[a]	Asteroid	Classification[b]	Possible Metallic Fe Content[c] (WT %)	Possible H_2O Content (WT %)
Carbonaceous chondrite (5)	887 Alinda	C30 carbonaceous chondrite	0·5	<2
	2100 Ra-Shalom	C3 carbonaceous chondrite	0·5	1-10
	1981 QA	C3 carbonaceous chondrite or E6 enstatite chondrite	0·19	0-10
Ordinary chondrite (81)	(1981QA)	(See above)		
	1862 Apollo	LL4 chondrite	0.3-3	'~0
	1980 AA	Shocked chondrite, dark	4-25	0.
Achondrite (9)	1915 Quetzlcoat[b]	Diogenite achondrite	1	0
Stony-iron	None identified[d]			
Iron (4)	None identified[d]			

[a] W. K. Hartmann, (1983). *Moons and Planets,* 2d ed. Based on tabulations by Scott (1978) and Dodd (1981). Note: Percentage of crumbly carbonaceous chondrites may be much higher in space, and percentage of other types correspondingly lower in space.
[b] Lucy-Ann McFadden, (1983), Spectral Reflectance of Near-Earth Asteroids: Implications for Composition, Origin, and Evolution, Ph.D. Diss. (Honolulu: University of Hawaii).
[c] Estimated from data on these meteorite classes, listed by J. Wasson, (1974), *Meteorites* (New York: Springer-Verlag).
[d] Because iron spectrum is relatively featureless, these are hard to confirm.

two were identified as carbonaceous chondritic and a third as either carbonaceous or ordinary chondritic. The others included two ordinary chondritic and one achondritic type. These data again suggest that a larger proportion of carbonaceous materials may be encountered in space than is represented in their 5 percent meteorite fall rate. This finding would imply that water and other volatiles may be more common in space than otherwise assumed. Water of hydration could easily be driven out of carbonaceous chondrite materials.

Two possible interpretations attach to the unidentified asteroid types found by McFadden. They may be essentially a known meteorite type, such as LL chondrites, whose spectra are superficially modified by effects of particle size, minor minerals, or some similar effect; or they may be fundamentally new types, not yet sampled in the meteorite collections. New types of meteorites are sporadically being identified in the terrestrial sample, especially the Antarctic collection.* We can expect at least as much variety among Apollo asteroids as among known meteorites.

If we assume that 60 percent of the Earth-approaching asteroidal material is carbonaceous chondritic to match its dominance in the lunar soil and redistribute the other types in proportion to the remaining known meteorite type abundances, we would encounter the type distribution shown in table 2.2 among Earth-approaching asteroids.

TABLE 2.2   Type Distribution Among Earth-Approaching Asteroids

Meteorite Type	Percent	Materials of Interest
Carbonaceous chondrite asteroids (including some burnt-out comets?)	60	Soil, water of hydration, organic compounds, possible ice
Ordinary chondrite asteroids	34	Silicate materials, scattered metal flecks, solar wind hydrogen
Achondrite asteroids	4	Silicate lava-like rocky material, solar wind hydrogen
Stony-iron asteroids	0.5	Silicates and large nodules of nickel-iron
Iron asteroids	1.5	Pure nickel-iron alloy
	100	

Anticipated Principal Resources and Extraction Processes

Based on the preceding discussion, we can tabulate the following resources of economic and life-support interest that may be encountered in asteroids.

---

*The science of meteoritics has taken giant strides in recent years with the discovery of fields of pristine meteorites eroded by wind from the Antarctic ice sheet.

Water ranges from about 1 to 22 percent by weight in carbonaceous chondrites and could be driven off by mild heating. It is plausible that there are ice deposits in deep interiors of some carbonaceous objects. Salts derived during ice melting at the time of comet activity might eventually seal pore spaces in the outer volume and allow ice in the core to be preserved if the "asteroid" does not spend too much time near the sun.

Nickel-iron alloy approaches 100 percent in iron objects and roughly 50 percent in stony-irons, but these may be hard to identify spectroscopically. Because of the lower strength of the stony component, a thin regolith of pulverized stone might mask the metal phase, possibly requiring *in situ* studies to reveal the metal content. Ordinary chondrite asteroids could also have large amounts of metallic iron, ranging around 15-25 percent by weight in Enstatite and H-group chondrites, and 0.3-15 percent in LL- and L-group chondrites. Carbonaceous chondrites have 0-8 percent metallic iron. In chondrites this iron appears in flecks and might be scattered through regolith material. Magnetic rakes could "harvest" these materials.

Platinum-group metals are dissolved among metallic phase grains, especially in ordinary chondrites. Unlike other resources, these have commercial values of thousands of dollars per kilogram. As emphasized by John Lewis, the United States is highly dependent on other countries (U.S.S.R. and neutral nations in some cases) for these materials. They are therefore especially attractive as candidates for refining and returning to Earth. Lewis and Meinel assert that "all common classes of meteorites contain higher concentration of platinum-group metals than the richest ore bodies in Earth's crust." Lewis concludes that these materials "make or break" space mining because of their economic and strategic values, and he advocates a carbonyl extraction process for refining these materials from asteroids. This process involves CO gas being passed through the material at 1 to 1,000 atm pressure and 100° temperature. Under such conditions, iron, nickel, and several other metals react to form gaseous carbonyl compounds. For example,

$$Ni + 4CO \rightarrow Ni(CO)_4.$$

These gases may then be condensed and the metals deposited. Under selected conditions, Fe and Ni can remain volatilized, whereas cobalt and platinum metals are left as a magnetic dust. Lewis states that this process has been used commercially on Earth and is readily adaptable to space.

Hydrogen implanted from the solar wind into regolith grains can be easily driven off by mild heating of regoliths.

Hydrocarbons among the carbonaceous minerals in carbonaceous chondrites will provide a wide range of organics, carbon, hydrogen, and so on and may be valuable for various future space industries.

Oxygen and sulfur-bearing minerals are abundant. If abundant ices are found in near-Earth space, electrolysis can produce oxygen. However, as early as 1965, Rosenberg, Guter, and Miller described an experimental system for the manufacture of oxygen from lunar materials. Their technique, developed at Aerojet-General Corporation, produces CO and then H_2O from silicate minerals. It involved the following reactions:

$$MgSiO_3 + 2CH_4 \xrightarrow{1800°C} 2CO + 4H_2 + Si + MgO$$

$$2CO + 6H_2 \xrightarrow[\text{Ni catalyst}]{250°C} 2CH_4 + 2H_2O$$

$$2H_2O \xrightarrow[\text{Electrolysis}]{75°C} 2H_2 + O_2$$

The carbon monoxide (CO) would be of obvious use in the carbonyl process for extracting metals.

Dirt provides building and shielding materials. This can include leftover material after volatiles, metals, and other valuables have been removed. There may be strong military interest in shielding materials, especially conducting materials that could screen against electromagnetic pulses. As early as 1965 Kopecky and Voldan discussed the possibility of fabricating useful objects from cast basalt for possible use in lunar exploration. They traced the roots of this industry in Europe back to at least 1852. Pipes, titles, and other structures of cast basalt are currently being manufactured, and this technology may be very applicable to asteroidal or lunar materials.

Energy is an additional space resource, easily harvested by building solar collectors whose sizes are virtually unlimited because of the zero-gravity, low-stress environment.

The existence of regoliths and asteroid bulk strength are important factors in the potential use of all these resources. Regoliths allow easy access to prepulverized materials. Some carbonaceous and ordinary chondrites (especially shocked ones) may be virtually rubble piles easy to disassemble. But irons may be very difficult to utilize because of high tensile strength. These questions require further basic research on asteroids.

A Principle for Space Resource Utilization

We can identify two extreme opinions about utilizing space resources. One is held by certain advocates of "free-enterprise" space development, including some members of the L5 Society and other space advocacy groups who visualize raising private capital, forming space industrialization corporations/conglomerates/government agencies, and mining space resources for the direct benefit of their investors. In their view the group (nation, corporation, or whatever) with the initiative and ability to begin using asteroid resources should be able to use the benefits as it sees fit.

At the other extreme, people argue that all space resources are already the common heritage of humanity and therefore must be developed only by joint consent of all nations. This view is commonly encountered among third-world nations whose leaders are concerned that space resources would simply aggravate the growing disparity between developed and underdeveloped countries if acquired by first-world Western conglomerates or second-world Soviet agencies. The free-enterprise extremists argue that any decision-making involvement by third-world governments would so complicate space development that no investors would be found to participate; hence, the process would literally never get off the ground. The shared-heritage extremists argued for a space development governing body in the United Nations during drafting of the Moon Treaty, but advocates of the free-enterprise position subdued the language in the treaty draft and blocked its ratification by the United States. The current draft of the Moon Treaty is reported to contain the doctrine that space resources are a common heritage of humanity, but does not define how this doctrine should be applied in practice.

These considerations lead to a principle that I suggest be applied to all steps in the study and utilization of space resources:

> Space exploration and development should be done in such a way as to reduce, not aggravate, tensions on Earth.

This principle is important because if space resources are as abundant as we can now postulate, they offer significant opportunity to either reduce or aggravate world tensions. If new raw materials and energy can be poured into the world economy from space, resource differences between nations could be reduced as a factor in world instability. By processing new materials in space, polluting by-products now being dumped into the biosphere can be reduced. Some systems of energy harvesting could allow nations to become energy independent. Wealth concentrations (for example, oil deposits) can be reduced as an aggravating factor in world affairs. We can accept the free-enterprise extremists' argument that if space utilization becomes bogged down by multinational bureaucracies, we may lose any chance to benefit the world by pursuing space resources. We can also accept the shared-heritage extremists' argument that if space re-sources are used in a way that only improves the living standard among the already wealthy groups of the world, this will increase economic disparities and clearly lead to a more unstable world. It may contribute to reasons why third-world political groups increasingly resort to violence to reduce perceived economic differences.

We need creative, cooperative thinking among nations to find ways of taking advantage of this opportunity to create a more stable global and interplanetary society (which can serve as the bases for still more extraordinary interstellar ventures), rather than going down the road of increased struggle to control the resources.

For the present the United States appears in a leading position to take the initiative in space development. We should therefore make well-publicized gestures toward applying the benefits to improve world stability and thus set precedents in this direction. For example, we might voluntarily pay for commercial rights to explore or exploit certain asteroids or portions of them. The rights might be sold by an international body, and the proceeds could be put into a global fund like the World Bank. This fund could be dedicated to projects that would allow third-world development and promote energy and re-source independence among nations. A second alternative is that we could voluntarily donate a fraction of the value of acquired or returned resources to the World Bank or some similar agency devoted to third-world development. This donation would be a sort of tax on all proceeds from space resources. Through such a tax the United States would establish a fruitful precedent for interpreting the so-far undefined common-heritage doctrine. It would still allow the profit incentive to work, but would co-opt Soviet or other accusations of rapacious self-interest.

More important, we would establish an exciting rationale for United States space exploration: stepping out from Earth to acquire and use space resources for the benefit of the whole world economy. We would thereby serve not just our short-term good but also our long-term good.

We can anticipate that the abundance of resources on low-gravity objects in space may allow a vigorous interplanetary civilization and economy to develop, based on transport of many materials between worlds. By gradually displacing polluting processes of mining and manufacture into space, this economy may even allow Earth to begin to recover its natural state, providing a cleaner and more cheerful environment for those who live here. But this can happen only if our generation ensures that the first steps are taken in a way that defuses current tensions rather than aggravating them.

References

Alfvén, H., and G. Arrhenius. 1971. Arguments for a Mission to an Asteroid. In T. Gehrels, ed., *Physical Studies of Minor Planets* NASA SP-267, Washington, D.C.

Anders, E. 1971. Reasons for Not Having an Early Asteroid Mission. In T. Gehrels, ed., *Physical Studies of Minor Planets*. NASA SP-267, Washington, D.C.

Chapman, C., J. Williams, and W. Hartmann. 1978. The Asteroids. *Ann. Rev. Astron. Astrophys.* 16: 33.

Cruikshank D., and W. Hartmann. 1983. The Meteorite-Asteroid Connection: Two Olivine-Rich Asteroids. *Science* 233: 281-283.

Gaffey, M., and T. McCord. 1979. Mineralogical and Petrological Characterizations of Asteroid Surface Materials. In T. Gehrels, ed., *Asteroids*. University of Arizona Press, Tucson.

Gradie, J., and J. Veverka. 1980. The Composition of Trojan Asteroids, *Nature* 283: 840.

Hartmann, W., D. Cruikshank, and J. Degewij. 1982. Remote Comets and Related Bodies: VJHK Colorimetry and Surface Materials. *Icarus* 52: 377.

Kopecky, L., and J. Voldan. 1965. The Cast Basalt Industry. In Geological Problems in Lunar Research. *Ann. NY Acad. Sciences* 123: 1086.

Lebofsky, L. 1981. The 1.7 to 4.2 µm Spectrum of Asteroid 1 Ceres: Evidence for Clay Minerals and Ice. *Icarus* 48: 453.

Lewis, J. S. 1974. The Temperature Gradient in the Solar Nebula. *Science* 186: 440.

Lewis, J., and C. Meinel. 1983. Asteroid Mining and Space Bunkers. Defense Science 2000+.

McFadden, L. 1983. Spectral Reflectance of Near-Earth Asteroids: Implications for Composition, Origin, and Evolution. Ph.D. Diss., University of Hawaii.

Rosenberg, S., G. Guter, and F. Miller. 1965. Manufacture of Oxygen from Lunar Materials. In Geological Problems in Lunar Research. *Ann. NY Acad. Sciences* 123: 1106.

❋3

G. D. Brin

ROCS' EGGS AND SPIDER WEBS: THE FIRST HARD STEPS TOWARD BUILDING STARSHIPS

It now appears increasingly feasible that we may be capable of launching some variety of interstellar expedition within a century or two. Projecting known physics into the engineering of the future lets us see that such voyages are probably neither too burdensome nor too impractical for an advanced society to initiate. These extrapolations have led to conferences such as this one, dealing with the likely modes and consequences of human interstellar migrations.

We must recall, however, that all projections, no matter how compelling, do not necessarily come true. All reasonable scenarios involving dispersal of human beings to the stars depend upon one critical first step—the industrialization of the Solar System.

For example, vessels suitable for interstellar voyages will undoubtedly be very large structures crafted from refined lunar or asteroidal materials. Such vehicles can only be constructed after much has been learned about macroengineering in local space. The multi-generation (or hibernation) life-support systems required will be based on lessons learned while maintaining isolated outposts off Earth over long periods.

Clearly, we will not create a major infrastructure outside of Earth's atmosphere with interstellar migration foremost in mind. The long-range vision of interstellar transport will probably have little influence over decisions to build major industrial facilities (and perhaps cities) within the Solar System.

Rather, the primary goal will be wealth. We will occupy the Solar System partly for adventure, partly for science, but mostly to get rich. Earth-returnable space resources, such as platinum group elements, are a very small part of the potential profit. Most of the repayment on investment will come from industry in orbit, using lunar and asteroidal materials along with copious solar power.

Removal of polluting heavy industry from Earth's biosphere, many-orders leveraging of biotechnology, low-gravity materials technologies, life extension, power transmission, and tourism are only a few additional possibilities. David Criswell's paper for this conference lays out a number of the opportunities, demonstrating how human per capita disposable income might expand tremendously soon after space industry is fully under way.

Once such an industrial base is fully established, the step from self-supporting factory-colonies to starships might be considered rather straightforward. Interstellar undertakings will begin as spinoffs from tremendous works within the Solar System.

On paper then the stages seem direct and clear—the future bright. Yet as great as these opportunities are, there is a strong possibility that they will not come to fruition. If, for near-term political reasons, the industrialization of space does not commence soon—and show rather immediate payoffs—then it is conceivable that exhaustion of earthly resources may cause a growing public insularity and an unwillingness to invest in "visionary" projects above Earth.

This threshold effect—in which opportunity sometimes seems to present itself only for a narrow period and then vanish—is familiar to us in our daily lives. There is every reason to believe that it holds for nations and peoples, as well as individuals.

(Some even propose that this choice is common to assertive extraterrestrial species who reach our level of development and further that most such extraterrestrials fail to take advantage of the brief opportunity when it occurs. This pattern might help explain the apparent absence of starfarers today, in spite of the apparent ease of such travel on paper.)

We seem to be approaching a "window of opportunity," through which this generation must pass. During this period ways must be found to make the occupation of space inexpensive and profitable enough—early enough—to attract continued and growing investment.

Particularly, it is clear that the profit potential from space must be demonstrated early, not on promises of a fifteen- to thirty-year return. Resources in Earth orbit must rise to a critical level before space industries will even begin tapping vast lunar and asteroidal reservoirs.

If what we have just asserted is true, then one cannot claim "manifest destiny"—that we will become starfarers as a natural evolution of current trends. Success will depend on our choice of near-term investment strategies—and in no small part on national and international will. The process will be helped substantially if imaginative techniques and resources are brought into play at the right time.

Some innovative new ideas have appeared that apparently offer just the bootstrap needed to lift us through the window of opportunity ahead. Fascinating possibilities include electrodynamic Lorentz engines in Earth orbit, which would use solar power but require no rocket propellant at all; solar sails for asteroid sample ore return missions; sophisticated robotics techniques allowing removal of expensive manned systems from mission profiles; and unconventional propulsive systems.

Of course, many of these concepts are merely optimistic glimmers at the present. Others appear to offer promising ways to dramatically multiply our capabilities. Some of these new ideas may well prove the difference between tepid, desultory investment in space, going nowhere, and the kind of major commitment that will lead almost assuredly to interstellar migration.

It might be instructive to look at one example, a resource that until only recently was virtually unnoticed, yet may end up carrying an infant space industry across the barren years until the lunar and asteroid mines are open and running. This is the space shuttle external tank, which towers like an eleven story building over the attached orbiter on the launch pad and provides 775 tons of liquid hydrogen and oxygen propellants to the shuttle main engines during ascent from Earth. The only disposable part of the space transportation system (STS), this mammoth structure is currently discarded every launch, jettisoned into the Indian Ocean. Yet recent studies have shown that shuttle tanks can easily be brought all the way into orbit, at almost negligible cost.

We shall examine some of the uses to which this resource might be applied. If it can offer the feedstock needed to make space industry profitable in the near term, then the tanks may be the first step on the road to a starship factory. It can be argued that the external tanks, or ETs, may be the eggs from which the great system and interstellar birds of space will hatch.

Shuttle Tanks as a Developing Space Resource

In figure 3.1 we see a broken schematic of the external tank (ET). It consists primarily of two large chambers connected by an intertank region, along with plumbing, valves, and a massive steel beam to which the shuttle solid rocket boosters (SRBs) are attached for the first two minutes after launch.

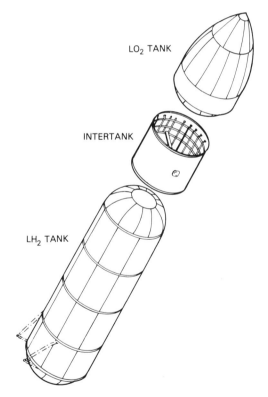

LO$_2$ TANK

INTERTANK

LH$_2$ TANK

Fig. 3.1. Structure of the Shuttle Main Tank. With a modest change in mission profiles, these could be placed in orbit for use in space development.

The larger chamber, containing liquid hydrogen, is more than 1500 cubic meters in volume. The liquid oxygen tank is more than one-third that size. Both are pressure-tested to more than 2 atmospheres, and both have detachable "manholes" at both ends—openings about the right size to be compatible with standard airlocks. The primary constituent of the ET is aluminum, with small portions of copper, zinc, steel, chromium, and polymers.

In 1981 the California Space Institute was commissioned to study applications of external tanks for the National Aeronautics and Space Administration. After several conferences involving representatives from industry, academia, the defense establishment, and various NASA centers, the following conclusions were reached:

1. Shuttle external tanks may be inserted into low earth orbit (LEO) at little or no appreciable cost—and possibly with some benefits—to the STS cargo bay lift budget. By retaining the tanks into orbit and counting their mass as cargo, we effectively double the maximum lift capability of the STS at no energy cost. The added possibility of an aft cargo compartment (ACC) offers a release from the volume-limited nature of most current shuttle cargos carried in the orbiter payload bay.

2. Perhaps even more valuable, the orbit-inserted tanks will carry angular momentum, which can be traded back and forth with other objects to great advantage. One way in which this can be done is by linking two massive objects together with a taut tether and using both objects as counterweights in either hanging or swinging systems in Earth's gravitational field.

Tethers allow the use of reaction mass other than rocket propellant. For instance, a tether made of conducting wire cutting through Earth's magnetic field can exchange momentum with Earth and its atmosphere; in other words, the system uses Earth itself as reaction mass. Tethers can also use physical objects as reaction mass; the bigger the mass, the better. External tanks are ideally suited for such a role.

The mass of external tanks may be key to creating flexible space station concepts, featuring detached modules, factories, observatories, cargo transfer stations, separated to prevent mutual interference yet all linked by a spiderweb of slender threads.

Momentum transfer techniques offer particular promise. At the very least, several methods appear to be available for orbit stabilization, maintenance, and/or disposal of ETs in or from LEO. If some of the more dramatic possibilities ever prove out, the thin traceries of linkage tethers may indeed become a web between the worlds.

3. The 31 tons dry weight of aluminum and other metals that constitute the bulk of the ET provide a stock of raw materials that

could be brought into LEO in no other cost-effective way in the intermediate future. The material is more easily recycled, melted, and formed than the task-dedicated machinery that will be carried up in the shuttle's cargo bay.

4. In addition to this 31 ton dry weight of metals and polymers, from 2 to 20 tons of residual cryogenic hydrogen and oxygen propellants can be recovered from an ET within the first hour after orbit insertion. NASA is already looking at these residuals for possible fueling of upper stages, powering fuel cells, or creating water for a wide variety of applications. Water appears to drive up to 40 percent of the lift budget for establishing and maintaining space station life-support systems. Bonus water from recovered ETs can open the way to dramatically larger and more capable manned facilities than those currently envisioned. It would also allow large-scale experiments in controlled ecological systems, a necessary step before we move on to deep space migrations.

5. The tanks themselves may prove valuable as rigid bodies; as large, airtight volumes (for habitats, shelters, warehouses, waste dumps, or farms); or for concealment or protection of sensitive assets.

A detailed examination of these resources, along with many of the uses to which they might be applied, is available in the proceedings of the California Space Institute Conference on external tanks. The ET appears to offer a large share of the basic ingredients required for extended stays in space—large airtight volumes already thermally insulated, having access to propellants, oxygen, and water.

Constraints

Any new idea can sound too good to be true. Enthusiasm must be tempered with a hard look at the basic factors that will limit the availability of a resource.

One fundamental limitation on the availability of tanks will be the number of STS flights. Still, at fifty launches per year ETs would represent approximately 2,000 tons a year of metals, ablatives, and volatiles. If these were sent into orbit, they would exceed by far the carriage envisioned via the shuttle cargo bay.

ETs will only be orbited if a number of political, safety, and economic considerations are satisfied.

1. Since the public relations disaster of Skylab, it is a political requirement that orbital stability and control must be provided, including both orbit lifetime extension (by tether orientation or other techniques) and deorbit and disposal capability.

2. Orbital insertion must not noticeably detract from already existing mission requirements for the STS.

3. Orbital inclinations and specific orbits of STS missions will mean that, even allowing for space stations, not all ETs will be collectible at warehousing points. Some will enter orbits that have little likelihood of repetition or reuse. These must be disposed of.

4. Retrieval, link-up, and manipulation at the site of use will all require development of new technologies that seem straightforward but nevertheless will take time and investment to perfect.

5. Any applications involving use as habitats will require intensive study before approval for "man rating." Also needed will be a breakthrough in psychological resistance to reusing materials instead of fabricating totally new forms and structures for each and every new purpose in space.

Conclusion

Between 1983 and 1994 NASA and the Department of Defense plan more than 300 STS missions carrying more than 7,000 tons of cargo into LEO in the shuttle bay. This will probably exceed amounts lifted by all other powers combined. The price for delivering this amount of mass into orbit is now approximately $2,500/kg ($1,140/lb). This is rather expensive for initiating space industrialization. It exceeds cost of fabrication of almost any product imaginable on Earth (with the exception, perhaps, of certain types of pharmaceuticals).

Furthermore, most of this cargo will be scattered into dispersed orbits. It will consist primarily of dedicated electronics and support equipment—in satellites of various types. Even if the material could be regathered and collected once it finishes its primary task, it will be of little recycling potential. However, by 1994, 10,000 tons of external tanks and about 5,000 tons of recoverable propellants will also reach about 98 percent of orbital velocity. At $2,500/kg these materials represent a $40 billion asset that can easily be added to the stock in orbit.

These 15,000 tons can provide the bootstrap mass required to establish the first major industrial facilities in space, as well as the feedstock from which biological life-support systems can be developed. Eventually, materials from ETs can be used to build facilities to mine Earth-passing asteroids or the lunar surface and get the true occupation of the Solar System under way.

There are many new ideas whose implementation may furnish the key to space. We chose to examine reuse of STS ETs here as an example of the innovative thinking that is going to be required. Putting these mammoth aluminum eggs to use—instead of continuing to dump them in the ocean—will be allegorically very much like recycling soft-drink cans: a polite, sensible, and profitable gesture on our way to greater things.

Every road has its beginning, and often it is the first few steps that are the hardest. It might be very good symbolism—as well as the height of practicality—if recycling a throwaway resource enables us to pass through the narrow window of opportunity from a parochial, Earth-centered economy to one based on ambitious and profitable space industry, leading eventually to the stars.

References

Bekey, Ivan. 1983. Tethers Open New Space Options. *Aeronautics and Astronautics*, April.

Carroll, Joseph. *On-Orbit Uses of Aluminum from the STS External Tank*. Calspace Report no. 83-05.

Forward, Robert. 1980. Interstellar Flight Systems. Preprint of a paper delivered at the annual meeting of the American Institute of Aeronautics and Astronautics, Baltimore, May 1983.

Forward, Robert D. 1983. *Alternate Propulsion Energy Sources*. Air Force Rocket Propulsion Laboratory Report TR-83-039.

Gustan, E., and T. Vinopal. 1981. *Controlled Ecological Life-Support System: Transportation Analysis*. Boeing Aerospace Company. NASA Contract Report no. 166420.

Lewis, J. 1983. Asteroidal Resources. Defense 2000, July.

Report on the Utilization of the External Tanks of the Space Transportation System: A workshop held at the University of California, August 1982. Calspace Report no. 83-01.

❋ 4

David R. Criswell

SOLAR SYSTEM INDUSTRIALIZATION: IMPLICATIONS FOR INTERSTELLAR MIGRATIONS

In my father's house there are many mansions.   If . . .

Industry: Earth to Space

Our intent is to pursue the implications of a major extrapolation of human industrialization into the Solar System. Extrapolations of a few long-term trends of human industrial activities and knowledge may provide insights into how other supposed agencies of organization (organic, inorganic, combinations . . . ?) elsewhere in the Galaxy or universe might affect their own locales. Although we hope these extrapolations might also provide wider physical starting points for discussions of the motivations for, support of, and inhibitions to migration of agencies of organization from one star system to another, the reader is reminded that many are very speculative. Others have pointed out that agencies that organize large sections of a star system, Galaxy, or cluster might produce effects detectable from Earth. We will explore a few newly recognized opportunities for and obstacles to detection. If organizing agencies have been operating for considerable periods of time in other regions of the universe, then their activities might be relevant to the development of our own space industrialization, as well as to our understanding of cosmology and the related sciences.

Figure 4.1 is a very simple representation of industrial development. Human beings are gradually increasing in their skills to manipulate matter and energy to build constructs or systems. We call the human produced results "cumulative controlled connectives," C+C+C, or CCC for short. The growth process is extremely interactive and complex. Early CCC advances were the discovery of fuels and fire, irrigationhuman minds and human groups. Some CCCs are transitory, like a radio signal, whereas others are relatively long lived, like a radio station or Roman aqueducts.

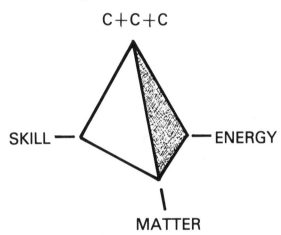

Fig. 4.1. Four interrelated factors of industrial growth. C+C+C means "cumulative controlled connectivities."

Approximately 400 years ago industrialization began in Great Britain. Human skill evolved to using wood and coal to operate on various ores to produce iron and other refined CCCs (engineering materials). The CCCs could then be used to make products for other applications (other types of CCCs) and to expand the basic energy gathering and mass manipulating systems (also CCCs). Pyramid power in figure 4.1 has underpinned the steady increase in human populations. Human skill is thereby increased, and the human developers and controllers have organized increased flows of energy, mass, and CCC.

Figure 4.2 presents the approximate increase in the human manipulation of nonrecoverable matter on a worldwide basis since the 1800s. Manipulation of the nonrecoverable material (sand and gravel) and energy resources (coal, oil) has been increasing at approximately 6 percent per year over the past 400 years. Also included as gifts of nature are the ores and other chemical concentrations used to obtain metals, chemicals for processing, and agricultural fertilizers. The major nonrecoverable elements manipulated in the mid-1970s by world industry are summarized in table 4.1. These nonrecoverable elements may be thought of as composing a synthetic molecule of industry, which can be referred to as demandite, a concept coined by Goeller and Weinberg.

Notice in table 4.1 that fuels (coal, oil, and the like) and heavy construction materials (sand and gravel) constitute the bulk of the nonrecoverable materials. As human industry moves into space, it will be possible to substitute solar power for organic fuels. This will eliminate the gathering of fossil or nuclear fuels and problems of disposal of burned products. In addition, very low mass systems in space will convert solar energy into other useful forms of power. Table 4.2 gives the chemical composition on a weight fraction basis of terrestrial demandite and of nonfuel or space demandite. Notice that an ore as poor as one known lunar soil can supply more than 92 percent of the necessary elements, by concentrating them by a factor of ten or less from the bulk soil. Criswell and Waldron showed that substitution of materials will increase the percentage of supply even higher. In figure 4.2 the industrial matter cycles are of the order of the natural biosphere cycles. Cellulose (table 4.3) can be considered the synthetic molecule of the biosphere. Averaged over the entire Earth, the living biosphere weighs about 0.5 g/cm^2 or 30 trillion (3×10^{13}) tons.

By the year 2000 there will be approximately 6 billion people. Assuming an average weight per person of 60 kg, the modern human population will weigh about 360 million tons or twelve parts in a million of the total biosphere. In the earlier hunter-gatherer stage, it is estimated there were only 30 million people, 200 times less than today and 10 times less than in the later agricultural period, which began about 10,000 years ago and is ending this century. It is not clear that human beings could constitute as much as 100 parts in a million of the biosphere on a long-term basis. We now live in an artificial world on Earth in which our human population is sustained by an

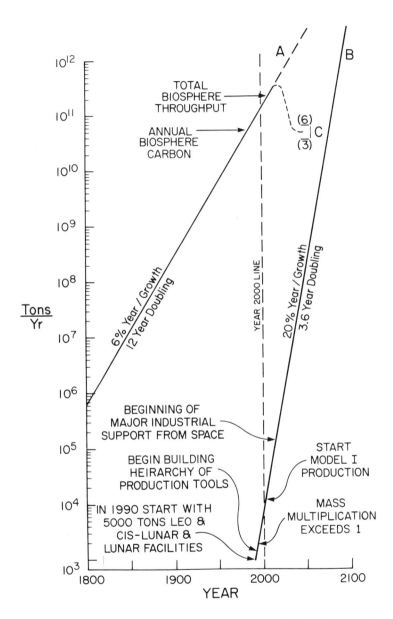

Fig. 4.2. The annual production of demandite on Earth (A) over the industrial era and the projected production in the early phases of space development (B).

TABLE 4.1   Demandite

Use	Molecular Composition Weight fraction	Components	Molecular fraction
Fuel	0.46	Hydrocarbon (CH_2 to C_6H_{14})	0.8022
Building	0.45	Silica (SiO_2)	0.1115
		Calcium carbonate ($CaCO_3$)	0.0453
Metals	0.06	Iron	0.0110
		Aluminum	0.0011
		Magnesium	0.0004
		Cu,Zn,Pb	0.0004
		Minor metals	0.0008
Agriculture	0.01	Nitrogen	0.0076
		Phosphorus	0.0008
		Potassium	0.0007
Others	0.02	Oxygen	0.0053
		Sodium	0.0053
		Chlorine	0.0053
		Sulfur	0.0023

immense annual flow of demandite, stored energy, and growing application of human skills.

Except for the production of fresh rainwater, sunlight for illumination and outdoor heating, and oxygen recycling (trees in the Amazonian forests—the primary repository of biomass), humans are living in a system that is decoupled from the biosphere. Human-controlled reclamation of water is rapidly increasing worldwide. Coal is the primary, rapidly depleting, and pervasively polluting fossil resource of ancient and inefficiently converted solar energy.

Industry must make fundamental changes as it evolves on Earth and moves into space. Solar energy can be efficiently obtained directly in space or on the moon and can even be provided to users back on Earth. There are very basic advantages. Far less reorganized matter or CCC (fig. 4.1) will be required (tens of tons per megawatt) to acquire and distribute the primary energy than with present or alternative Earth systems (100 to 10,000 tons per megawatt). Space solar power systems can grow rapidly in capacity (tens of days energy payback times). They will require very little maintenance and will not require

TABLE 4.2 Demandite and Nonfuel Demandite Compared to Elemental Distribution From One Moon Site

		Weight Fractions		
Element	(1) Demandite	(2) Nonfuel demandite	Apollo 15 mare (3) (low titanium)	(2)/(3) Enhancement required[a]
Carbon	0.3551	0.0574	0.000095	604.0[d]
Hydrogen	0.0763	0.0025[c]	0.00007	35.71[d]
Silicon	0.1355	0.2444	0.2158	1.13
Oxygen	0.2217	0.4547	0.4130	1.10
Iron	0.0797	0.0479	0.1535	0.31
Aluminum	0.0039	0.0023	0.0546	0.042
Magnesium	0.0013	0.0017	0.0681	0.025
Copper, zinc and lead	0.0033	0.0020	0.000022	90.0[d]
Manganese[b]	0.0057	0.0030	0.0189	0.16
Calcium	0.0473	0.1417	0.0696	2.0
Sodium	0.0158	0.0095	0.0023	4.1
Sulfur	0.0095	0.0058	0.0006	9.7
Potassium	0.0036	0.0021	0.0008	2.6
Phosphorus	0.0032	0.0019	0.0005	3.8
Chlorine	0.0244	0.0147	0.0000076	1934.0[d]
Nitrogen	0.0138	0.0083	0.00008	103.0[2]
Total	1.0001	0.9999		---

[a] Required to meet nonfuel demandite fraction.
[b] Manganese, titanium, chromium, barium, fluorine, nickel, argon, tin, bromide, zirconium, and boron.
[c] For use in plastics; does not include water.
[d] Times 10 enhanced lunar soils 8% deficient in this element.

the continuous movement and dispersal of extremely large quantities of mass (as with the burning of coal or oil on Earth). Present day world energy needs are the order of 5,000 GW (GW is 10^9 watts) and can be met by the fusion burning of 1.78 tons per year of hydrogen and other nuclei in the interior of the sun. Energy needs of a materials- and energy-prosperous world of 6 billion people using 20,000 GW could be met by intercepting in space the energy flow of the burning of 7.12 tons per year of stellar mass. Energy could be brought to Earth directly via microwave beams and embodied in products or refined materials. Notice in table 4.2 that the soils of the dry, barren moon contain most of the chemical elements obtained by present-day Earth industry from nonrecoverable resources. It is clearly possible to devise processing systems that can obtain the primary industrial materials of society

TABLE 4.3 Life or Cellulose

Element	H	O	C	N	P	S
Molecular fraction	2860	1428	1428	16	1.8	1
Weight fraction	0.066	0.530	0.397	0.005	0.001	0.001

from both the unenriched resources of the moon and the common crustal composition of Earth. Closed-loop chemical and physical systems are possible that can use immediately obtained solar energy as the main waste output and produce upgraded engineering materials from common resources with virtually no secondary materials pollutants. Compounds and elements recycled by solar energy on Earth (H, C, H_2O, and so on) can and will also be accumulated and recycled in space systems. Access to space will greatly speed up growth of our energy supplies and access to useful matter in space and on Earth.

We are in the beginning phase of fundamental changes in skill. Computers, electronic communication systems, investigations into artificial or machine intelligence, and advanced manufacturing and robotics are rapidly advancing the time when manipulative skills can be stored in manufactured systems. Such advances are decoupling skill from the number and level of health and training of human beings. We see this change in the United States in the steady decrease of the fraction of our work force required for the basic and manufacturing industries. Conceivably, the fractional population of people in industry will decrease to the few percent level or less as has happened in agriculture. A similar trend is apparent in white collar or middle management portions of the economy. Major social adjustments must occur. The complement of this is that progressively smaller groups of people will be needed to organize, initiate, and complete progressively larger programs. This is relevant to space.

Space industrialization will use the leading-edge technologies or skills. A few people on the moon and in cis-lunar space will be able to initiate, direct, and maintain rapid increases in the growth of materials industries off Earth. There should be powerful terrestrial incentives to support this growth. In view of the fundamental advantages of the access to and growth of energy, mass, and skills off Earth, there should be a rapid production of support facilities and systems (CCC) off Earth. It seems reasonable to expect that the manipulation of matter can grow at appreciably higher rates in space than on Earth. It seems reasonable to expect that space-using societies will encourage rather than resist large-scale projects in space. Needs for large facilities and systems will be great. We very roughly estimate that a 20 percent per year rate of increase is possible. Such a growth rate, which is not really critical to the arguments in the following section, implies the potential for amazing developments relevant to the permanent human occupation of space.

Figure 4.2 illustrates the results of creating on the moon in the year 1990 a small (5,000 tons) initial solar power system, gravel pit, local manufacturing complex, and system for transporting materials from the moon to space. A 5,000-ton installation could grow very quickly from a much smaller initial seed base (say, 50 to 500 tons) placed on the moon a year or so earlier. The starting date is not critical, except to the dominant sponsors. Only a few flights of the space shuttle to low earth orbit (LEO) would be required to start up the operation. Rockets using lunar derived propellants could efficiently convey people and machines to the moon. Lunar materials could be used in many ways to expedite large-capacity transportation from LEO to deep space at low unit ($/kg) cost; support growing industries in LEO, cis-lunar space, and on the moon; and provide large quantities of materials to protect and sustain human life in space and relieve human-industrial pressures on the biosphere of Earth. The 20 percent per year line in figure 4.2 indicates that the mass of the initial lunar system would be doubled by approximately 1995 at about the time more mass is manipulated to make products in space than is supplied from Earth. By the year 2000 adequate productive capacity would exist off Earth to begin manufacture of modest permanent habitats approximately the mass of a nuclear aircraft carrier; the basis for this would be O'Neill's Model I habitat. Early in the twenty-first century, space derived material products would begin to be significant fractions of the terrestrial economy. There would be widespread human access to space. We should remember that the fuel required by a modern jetliner to fly halfway around the world is about the same as to put the same payload into low Earth orbit, a fact first noted by Hunter and Fellenz. Development of routine semiballistic commercial flight around the world would not only bring all parts of the world within one hour travel time but would also bring Earth orbit to within twenty minutes for most travelers at virtually no extra cost. It is reasonable to expect that tens of millions of people a year may travel close to space by the early twenty-first century. Where might growth in space, easy access to space, and the tying of the economy of Earth to space activities lead?

Many authors have projected the conversion of the moon, asteroids, and even the planets into habitable human abodes. In a seemingly more artificial but probably more realizable vein, others project the creation of permanent human habitation of large cylinders, spheres, or other constructs (see, for instance, Olaf Stapeldon's discussion from the 1930s). The very words *habitats, worldlets, colonies, space stations, spaceships,* and so on, evoke in many people feelings of restraint, dependence, fragility, and even boredom. Future space dwellers will very likely find it extremely hard to imagine the emotional and intellectual wellsprings of such feelings. They will consider their space homes or *s'homes* (first coining) as natural, highly variegated, and supportive places. Feelings of safety will reasonably spring from the certain knowledge that their advanced technologies constructed, operated, and can constantly refine their places and build new ones. Starting from scratch is probably the best way to explore, construct, and understand systems as complicated as biospheres. Early space colonies supporting a few hundred to a few thousand people would be appropriately sized test tubes within which to develop the biosphere sciences. Results and progress can be far firmer than attempting to work only within the almost alchemical contexts of the immense and changeable biosphere of Earth. Long-term results would probably be obtained at less costs than on Earth. S'homes could be used to establish permanent habitations throughout the Solar System. Travel between such cooperative centers would become routine and inexpensive.

Macro s'homes could be constructed primarily from lunar materials and placed in the two stable gravitational wells, named after the French mathematician Lagrange, 60° in front of (L4) and behind (L5) the moon in its orbit about Earth. L4 and L5 are somewhat ellipsoidal volumes the order of 70,000 km in radius. We will approximate their usable projected surface area for intercepting solar energy as a circle 50,000 km in radius (8 billion km^2 or 8 billion GW of useful power) and the volume of each as a sphere (5 × 10^{14} km^3). It is argued that cylindrical s'homes, each scaled to accommodate approximately one million people, would have a mass the order of 100 million tons or about 100 tons per person. Other designs are more economic of mass (SP-413). Assuming each s'home dweller consumed 200 kw of solar energy, the L4 and L5 regions could accommodate 40 trillion (4 × 10^{13}) people with 4,000 trillion (4 × 10^{15}) tons of supporting structure. Figure 4.3 (the 20 percent per year projection) indicates such a scale of construction might be possible by 2115 assuming a ten-year average construction period. The total volume of L4 and L5 occupied would be 1.2 billion km^3 or only one-millionth of that available. Construction could greatly outpace the growth of human population, which could be expected to be about 240 billion (2.4 × 10^{14}) at a 3 percent per year growth rate. A 7 percent human growth rate would be required to provide 40 trillion people. By 2115 each of the L-volumes could support more individuals than the number of neutrons in an individual brain. This vast number of people could be in contact via laser communication with delay times of 0.2 seconds or less in a given L-volume and a second between Earth or the other volume. Frictionless and therefore extremely cheap transport of people and goods inside and between the Lagrange volumes would be

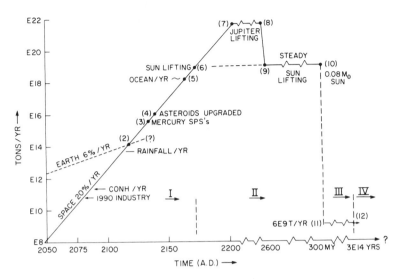

Fig. 4.3. The course of mass processing in the solar system before and during the period of stellar husbandry.

possible. Technological prowess of the two Lagrange volumes would be immense.

The cool and condensed resources of the Solar System are relatively small when faced by a technological society growing at a vigorous mass handling rate of 20 percent per year or even by a society that develops at a much slower rate, say, 1 percent per year. I first realized the finite limits of the Solar System while reading an article by Wetherill on the influx of new asteroids to the inner Solar System owing to the gravitation deflection of comets into tighter orbits about the sun. Approximately 100 billion tons per year of new working material was indicated. This was rather depressing. It is a level comparable with the present world production of primary materials. Would a growing space economy process major components of the cool resources of the Solar System and then run into materials limitations not significantly different in scale than those presently accessed on Earth? A certain irony of this perceived limit will be indicated later.

Figure 4.3 shows the implications of 20 percent per year growth for the twenty-first century and somewhat beyond. By 2080, less than one hundred years from now, space industry would be the scale of present world industry. It is not unreasonable to expect that terrestrial processing of primary or nonrenewable materials could stabilize at 1990s levels given access to space solar energy and products. However, there is no reason to restrict growth distant from Earth. Thus, by

2085 mass handling might be approximately the same as the through-put of the entire terrestrial biosphere. This implies that construction of total habitat surface areas the scale of the terrestrial biosphere every year would be possible. In the 2110s space production could exceed the recycling on Earth of rainwater from the oceans and a couple of years later, the total production of Earth, assuming the unlikely continued 6 percent growth rate of industry on Earth. Most of the asteroidal mass of the Solar System would be upgraded by A.D. 2140. At this point continued growth has been forecast to require the disassembly of the moons and planets. Dyson and others have proposed various schemes consistent with classical physics that would permit planetary upgrading. Jupiter-lifting (that is, taking the planet apart) is shown in figure 4.3 as starting in A.D. 2200 and being completed approximately 400 years later. The pitifully small influx of new cometary materials into the twenty-seventh century Solar System would be of no consequence for the support of such extensive habitable volumes as are possible by processing of debris, moons, and planets. Are we considering the appropriate end point for develop-ment of our local resources? Might not our proper focus for local husbandry be our sun? Our sun is our primary local energy source. It might also be our major source of materials. However, its primary utility might be to manage the even more precious commodity for our species—our duration in space-time.

Stellar Husbandry

Student: (suddenly alert) Professor? When did you say it will burn out?
Professor: Our sun? About 5 billion years from now.
Student: What a relief, I thought you said 5 million!

Line 1 in figure 4.4 depicts a well-known fact in astrophysics. The larger a star is, the faster it consumes its burnable fractions of hydrogen, helium, and other fusionable elements. Table 4.4 gives the relevant numerical values of power, mass, and so on, applicable to the sun. Sol has a projected life of 12 billion years. Sol pours out an immense stream of quality photons (1,000 to 50,000°K) from its 5,800° surface into the 3°K blackness of space.

It does this by fusion, converting primarily hydrogen to helium and thereby releasing energy. Approximately 0.7 percent of the mass of four hydrogen nuclei are released as energy to produce one helium nucleus. Another 0.1 percent of the remaining mass energy could be released by fusion burning helium to iron under more extreme

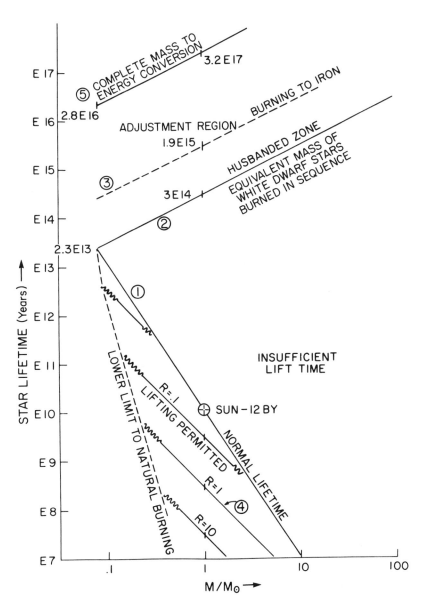

Fig. 4.4. Lifetime of natural and husbanded stars as functions of their mass.

TABLE 4.4   Natural and Husbanded Stars

Item	Sun	White Dwarf
Mass (tons)	1.99×10^{27}	1.6×10^{26}
Power output (Gigawatts or 1E9 watts)	3.9×10^{17}	1.6×10^{13}
Mass which is fusion burned per year (tons/year)	1.4×10^{14}	5.7×10^{9}
Normal lifetime (years)	1.2×10^{10}	2.3×10^{13}
Lifetime as successive white dwarfs (years)	3.0×10^{14}	----
Lifetime if all hydrogen could be burned to iron at white dwarf output (years)	1.9×10^{15}	----
Burntime at white dwarf power output if all mass could be converted to energy (years)	3.5×10^{17}	2.8×10^{16}
Star-lifting rate (tons/year)	6.5×10^{18}	7.0×10^{12}*
Number of people supported at 200 kw/person	1.8×10^{21}	8.0×10^{16}

*Mass would be circulated through the white dwarf at this rate to maintain the proper distribution of elements for hydrogen and helium fusion and extract the newly produced heavy elements at the rate of 5.7×10^{9} tons/year.

conditions. One percent of this energy comes out of the sun as high-speed solar wind (stellar composition) at the rate of 10 trillion tons/year. A star ten times as massive as the sun but of the same radius will consume the fusionable materials in its core much faster than the sun and explode in 120 million years.

Hydrogen balls of 0.08 solar masses are about the lower limit to a fusion-powered star. These small stars, referred to as white dwarfs, will burn about 2,000 times longer than Sol and are then expected to cool or evolve into brown dwarfs and eventually neutron stars depending on detailed conditions. They are described in the second column of table 4.4.

If Sol could be gently unwrapped of its outer layers and converted into white dwarf form, then the new dwarf would live 1,150 times the currently estimated age of the universe (20 billion years). The new white dwarf would partially burn its initial stores of hydrogen and helium. The unwrapping process will be referred to as star lifting. If the excessive materials taken from Sol were saved and then at appropriate times recompacted into a new white dwarf after the old one became too cool, then the civilization would be supplied with power for approximately 15,000 times the estimated age of the universe. Note line 2 in figure 4.4. Ages of ten-solar-mass stars could be extended to more than 3×10^{18} years.

If convection of fusionable materials from the inner to the outer portions of the star could be controlled, then longer lifetimes might be possible because the star could convert more of the original matter from hydrogen and helium to iron. Refer to item 6 in table 4.4 and line 3 in figure 4.4.

There is considerable interest in the possibility that black holes exist and can evolve from small stars. Astrophysicists have argued that energy could be extracted from a spinning black hole by dumping any matter (hydrogen, helium, or even iron, which will not fusion burn) along carefully chosen trajectories. The matter will release photons as it falls inward and into the black hole. Dicus and coworkers suggested that it seems possible to collect the released photons. Perhaps an advanced civilization capable of husbanding a star could convert the original stellar core or a derived core into a suitable black hole. If this could be done for Sol, then the energy release lifetime from the remaining mass would approach 16 million times the estimated age of the universe. This assumes that the Sol-derived black hole would output approximately 1.6×10^{13} GW. The cultured black hole could likely support 13 million times more people than on Earth now if each person used 200 kw of power. Refer to line 5 in figure 4.4 and item 7 in table 4.4.

A large fraction of the entire mass of the original star might become available in increasingly varied form over millions of billions of years. The civilization might direct its resources and time to converting matter into condensed forms or macromachines to manipulate all the forces of nature (electromagnetism, weak and strong nuclear forces, and gravity). We note, however, that there would be enormous penalties to ejecting major flows of matter from the star system on a continuous nonreciprocal or nonenergy conservative basis, such as in support of rocketlike, one-way interstellar flight on a large scale.

Our current technologies and energy needs cannot be expected to prompt us to seriously consider the prospects of dealing with management of the full solar resources. The industrially advanced nations are accused and self-accused of wasting the 20 kw (or 0.00002

GW)/person they consume to support their materially extravagant life-styles. However, we also live in an era of enormous waste of solar energy. Not only are we failing to use the 81,000,000 GW/person (averaged over everyone on Earth) of power the sun is currently sending irretrievably to cold deep space but we are also vigorously wasting the extremely inefficiently obtained power that has fallen on Earth over geologic time to produce our fossil fuels and power our inefficient biosphere. We currently burn several billion tons of coal, oil, and wood (fossil solar energy) a year worldwide (equivalent to 1.78 tons of fusion mass burning) to meet our meager energy needs, while we permit the sun to completely fusion burn nearly 136,500,000,000,000 tons of matter each year. We do have present-day limits to our technology, our will, and our immediate needs to tackle this stellar task. However, we can frame the basic challenges and begin imagining the general approaches to the conservation and upgrading of our stellar resources in the Solar System.

Star Lifting

Of course, the star must provide the energy for husbandry. Assume that 20 percent of the solar flux can be intercepted by solar power stations approximately 11 solar radii from the center of the star. Unmanned space probes are designed now that can operate this close to the sun. These power stations would provide about 50 percent of the intercepted energy to mass extractors. Power stations with giant collectors having an average thickness of 100 μm seem possible. The surface mass density is about ten times better than solar power satellites first proposed by Peter Glaser. The total mass would be 4×10^{17} tons. If materials were extracted from the planet Mercury, then approximately 0.01 percent of Mercury would be used. Referring to figure 4.3 this could start by A.D. 2140 for a 20 percent growth rate. Seemingly impossible schemes like the construction of Mercurian power stations may be possible by the application of exponentiating construction systems. In the first stage small machines are designed that use only local materials and solar power to make copies ₄of themselves. The copies make more copies and so on. Biological-like exponential growth appears possible, such as in simple bacterial cultures in the initial growth stage. When sufficient primary machines are accumulated (reproduced), then they are all directed to make the desired product (Mercurian power stations in this example). A 1982 NASA study and a best selling novel by Arthur C. Clarke have explored exponentiating systems in space applications.

Lifting solar plasma from the surface to infinity requires 1.9 × 10¹⁴ joules/ton (a joule is 4 × 10⁷ watt-seconds). Providing 10 percent of the solar flux for this task permits an uplift rate of 6.5 × 10¹⁸ tons/year. Approximately 300 million years would be required to uplift, cool, and store the materials for the transition to the white dwarf state (phase I). According to figure 4.3, star lifting could begin about A.D. 2170. Even the hypothetical prospects for such rapid technological growth might deeply frighten many people. We note that if the growth rate of Earth is 6 percent/year, then star lifting would not be possible until A.D. 2600 or by A.D. 5650 for a 1 percent rate of growth. These are both extremely short times by stellar standards. The periods are even relatively short compared with human history.

A general formula for star lifting (very approximate) can be given for stars on the main sequence. It is

$$T = \frac{300 \text{ million years}}{aM^2R}$$

where *M* is the stellar mass in units of the sun's mass, *R* is the stellar radius in units of the sun's radius, and *a* is the fractional conversion of the stellar luminosity to potential energy of the lifted materials. For a given *a*, lifting is quicker as *R* increases (lower potential energy difference) and the star gets more massive (more luminous energy to work with). Line 4 in figure 4.4 is the lifting time for stars of one solar mass as a function of radius (assuming *a* = 0.12). A star can become too cool to lift (down and to the left), too dense and too low power to lift (up and to the left), or simply take too long (up and to the right). Lifting times would be different for stars not on the main sequence. From the standpoint of macroengineering it is important to have the star provide the primary power to minimize mass of the lifting tools. These tools will be less massive if they do not have to produce their own power. Providing power sources would be a major difficulty when working presolar dust and gas clouds or extremely distended cool giant stars.

Figure 4.5 illustrates one approach to lifting of stellar materials. It would use extremely massive and powerful versions of the general types of particle beams systems discussed for space-based defense against ballistic missiles or in national accelerators facilities such as FermiLab or CERN. A large number of physically separate ion accelerators are placed in equatorial orbits about the sun. These ion accelerators fire two counterdirected beams of oppositely charged

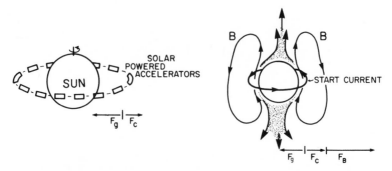

Fig. 4.5. Huff-n-Puff method for the gentle removal of mass from the Sun by polar ejection.

ions from one to the other and thereby around the sun. A powerful magnetic field is created, as shown in the second frame of figure 4.5. Once the ion beams are created, relatively little power will be required to maintain them or the magnetic field they induce. The electrically charged atoms in the atmosphere of the sun will follow this new large-scale magnetic field just as they do smaller magnetic fields of solar flares. However, the artificial magnetic field provides openings over the north and south poles of the sun through which plasma can escape. In effect, the artificial field forms a rocket nozzle over each pole of the sun. As strange as this concept sounds, the next step will seem even stranger. Few atoms in the atmosphere of the sun move fast enough to escape from the sun through the rocket nozzle. The upper atmosphere of the sun must be heated. This heating can be done with power from the solar power stations in orbit about the sun, which were first used to energize the ring currents and magnetic field. Microwaves, laser beams, neutral particle beams, or extremely high energy particles can be projected into the chamber of the solar rockets. Each chamber would be the volume between the polar region of the sun and the magnetic nozzle above that pole. Beginning evaluation of these heating techniques can draw on experience with heating plasmas in fusion power experiments on Earth.

Rockets work because the rocket nozzle forces the extremely hot exhaust gas to convert the random thermal motions of the burning chemicals into smooth flow out of the nozzle. Molecules moving downstream from the nozzle can actually constitute a relatively cool gas. The same can happen to the exhaust gases of the two solar rockets. At great distances from the sun the gases will stream outward as a relatively cool jet. Within the jet chemical reactions can occur to create solids and liquids of most of the elements except the dominate hydrogen and the noble gases. Excess hydrogen and the noble gases can be collected temporarily on the surfaces of dust grains. At a later time solar heat could be used to remove them and carefully introduce the gases into growing cryogenic gas spheres, which will eventually be self-contained by their own gravity. The dust can be recycled into the jet streams. There are likely to be other methods of collecting the solar gases. I will leave the conceiving and evaluating of additional concepts to others with more imagination than myself.[!!—eds.]

Other techniques should be considered for driving off the solar atmosphere. It may be that the accelerators (fig. 4.5) can be driven into oscillations toward and away from the sun by decreasing and increasing the ring currents at appropriate times. The accelerators would move around the sun slower than a natural planet at the same distance. The force of solar gravity (Fg in fig. 4.5) on the accelerators is balanced not only by the orbital centripetal (Fc) force but also by the outward force that the circulating ring currents (Fb) exert on the accelerators. The accelerators can push on the solar atmosphere via the variable magnetic field and create sound waves in the atmosphere to heat solar plasma in the rocket chambers above each pole. Natural solar wind is energized by sound waves or turbulence moving from the dense plasmas near the sun to the tenuous plasmas above the photosphere. We refer to magnetic pumping as the huff-and-puff method. It keeps the coldhearted wolf of entropy away from the front door of the sun for a few more years. Only relatively small flows of station-keeping mass would be required to maintain the huff-and-puff system. The star's gravity and energy and the symmetry of the mass expulsion permit this approach. After the accelerators are pushed outward, the ion beams are turned off, and the accelerators are allowed to fall inward to repeat the cycle. The sun currently loses approximately 4×10^{13} tons/year of ions via the solar wind. Star lifting would enhance the natural outflow of solar gases by a factor of approximately 100,000. Other stars are known that have such large outflow rates.

Figure 4.6 presents another approach, which mimics to some extent the explosion of mass from rotating stars with strong magnetic fields. Accelerators are placed in a common polar orbit about the star and slowly decelerated. At the proper moment, as they fall inward, counterdirected and oppositely charged ion beams are beamed between the accelerators creating current loops about the star. The magnetic field is not strong enough to stop the inward fall of the accelerators. At the appropriate moment, large ion rockets are started on each accelerator in such a profile of thrusting as to cause the plane of the accelerator orbits to begin to turn about the star as if that plane

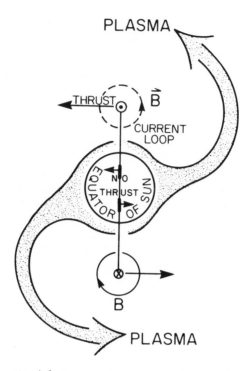

Fig. 4.6. Scheme for centripedal removal of
mass from the Sun.

pivoted about the spin axis of the star. The polar axis of the magnetic
dipole field lies in the plane of the equator of the star and forms two
channels through which plasma is slung into the ecliptic plane about
the star. This process can be continuous. However, it does entail the
constant production of extremely high energy ions for thrusting. The
thrusting ions need not be lost since they can be directed back into the
atmosphere of the sun and heat it. However, this heating may be at
considerable loss of thrust efficiency.

Features of Local Development

Husbanding of a star in this model divides into four general
periods. Refer to figure 4.3. First (phase I), there is the formation of
the star and planets, the evolution of life in the system (or its
introduction from outside at an appropriate time), and the development and industrialization of intelligent life on one of the planets or
large moons. Phase I would terminate with the beginning of star
lifting. From a distance stars in the prelifting phase would be virtually
indistinguishable from any others. There might be a very brief period

in which radio communications were detectable, but with improving technology general broadcasting would likely stop.

In phase II star lifting would occur. Immense mass flows would be directed from the star to space in highly symmetric patterns. Symmetric regions about the star might be strong emitters of thermal and radio emissions and of Doppler shifted emission lines of atomic recombinations. Of course, efficient engineering and operations would minimize these observables. I anticipate that the operators would seek to utilize the fully ionized nature of the gas to minimize the formation of complex or messy molecules to simplify later materials engineering.

Containment of the stellar hydrogen, helium, and other elements is possible. After all, the hydrogen and helium were previously held together by their own gravitational attraction against vigorous boiling by internal fusion burning in the sun. Hydrogen and helium could be reformed into gravitationally restrained spheres (small gas planets) at great distances from the star. Storage spheres of hydrogen and helium could be formed as very cold gases near the 3°K background temperature of the universe. However, reinjection of the stored hydrogen and helium into the husbanded white dwarf might produce observable effects. Conditions for extracting the stellar materials would change over time. As the star decreased in size, its rotation rate would likely be accelerated to aid in mass extraction. The extraction rate would decrease as the sun was guided toward its long-term husbanded state.

Phase II would be of a few million to a few hundred million years for main sequence stars. We assume cosmic abundance composition (by weight) for the lifted material, that is, 63 percent is hydrogen, 36 percent is helium, and 1.4 percent makes up the remainder (see table 4.5). Net new water (over and above the life and structural accumulations) could be produced initially (first 300 million years) at 1.6×10^{16} tons/year or a total of 4.8×10^{24} tons. It is rather difficult to keep this magnitude in context. So let us compare it to our Earthly experience. The oceans of Earth mass out at 1.42×10^{18}

TABLE 4.5 Summary of Elemental Availability for Cosmic Abundances

Wt. fraction of	H	He	C	N	O	P	S	Metals	Others	Sums
Cosmic abundance	0.63	0.36	2.8E-3	8.1E-4	6.7E-3	6.6E-6	3.3E-4	1.3E-3	2E-3	1.0
Life elements	3.6E-4	--	2.1E-3	2.8E-3	2.8E-3	6.6E-6	3.9E-6	--	--	5.3E-3
Plastics	1.0E-4	--	8.4E-4	--	--	--	--	--	--	9.4E-4
Oxide-based structurals	--	--	--	--	6.4E-4	--	--	1.3E-3	--	1.9E-3
Water	2.4E-4	--	--	--	2.2E-3	--	--	--	--	2.4E-3
Other uses	0.63	0.36	--	7.9E-4	--	--	3.3E-4	--	2E-3	≅3E-3

tons. Thus, we are discussing the creation of about 3.4 million new terrestrial oceans over the 300 million years. Jupiter and Saturn could provide an initial inventory of 80,600 oceans over the first 500 years of planetary mass lifting. Jupiter and Saturn are thought to resemble the sun in their compositions.

The other items in tables 4.5 and 4.6 are interesting. To a first approximation, life (cellulose as in trees) can be described by the distribution of chemical elements of cellulose as shown in table 4.3.

TABLE 4.6   Magnitudes of Mass Lifting (Sol) and Final Tonnages

Item	Mass lifting (tons/years)	Final (tons)
All mass	6.5E18	2.0E27
H	4.0E18	1.2E27
He	2.3E18	7.0E26
Life	3.4E16	1.1E25
"Plastics"	6.1E15	1.9E24
Oxide structurals	1.2E16	3.8E24
Water	1.6E16	4.8E24
Others	2.2E16	6.0E24

Approximately 800 times (by weight) the phosphorus mass of a sun could be converted into lifelike molecules, or 0.0053 of the weight fraction of the star. In making life molecules, phosphorous would be exhausted first. In phase II lifelike molecules (table 4.6) could be assembled at the rate of 3.4×10^{16} tons/year. On Earth the biosphere (mostly trees) has an average column density of approximately 0.5 g/cm^2. The radius of Earth is about 6,735 km and its total area 5.11×10^8 km^2. Assume biospheres constructed from solar elements have 1 g/cm^2 or 10,000 tons/km^2. Each year of sun lifting would enable the equilibrium production of 3.4×10^{12} km^2 of new biospheric area for a total of 1.1×10^{21} km^2 over time. This is 6,650 new areas/year of biospheres equivalent to the surface of Earth. In 300 million years about 2×10^{12} Earth area equivalents would be produced. Biospheres could initially outnumber oceans by a factor of about 633,000 to 1. As noted earlier, in the 1980s human beings constituted by weight 10 parts in 1 million of the biosphere. Using the life elements row of table 4.5, about 4×10^{11} tons of new materials would be available each year for living human beings to convert into additional humans. Sun lifting would be consistent with the birth of

7×10^{13} people every year (0.06 tons/person) or 2×10^{21} people at the end of phase II (fig. 4.5).

Solar energy would not necessarily be cheap in the economic sense if this many people came into existence on a steady basis. For example, if the population of humans off Earth grew at the rate of 1 percent per year starting in A.D. 2000, then by 5800 all the phosphorus available from our sun would be converted to 2×10^{26} humans. Decisions will always be with conscious entities. A fully husbanded star (careful cycling of hydrogen) would allow only 8W/person for a total population of 2×10^{21}. This is a far more severe constraint than on Earth where the sunlight is available for daytime lighting, driving weather, and recycling water and driving photosynthesis at the level of 32,400 kw/person (0.1 albedo) averaged over the surface of Earth. However, the volumes in space could be designed for maximum efficient use of energy flows in ways not possible on Earth. Of course, there is no reason to think that the humanlike population would be anywhere close to 2×10^{21} individuals. Assuming a per capita consumption of 200 kw, then a husbanded star could support 8×10^{16} people, or about 20 million times more than now on Earth. Each person could be supported in phase III by access to manipulation of 36 million tons of mass other than hydrogen and helium.

Using 10 tons/m^2 of surface area for materials of construction of s'homes (highly arbitrary for 300 million years from now), the structural metals, oxides, and plastics would provide 1 billion km^2/year of new living area. On completion of star lifting, there would be 5.7×10^{17} km^2 or a billion areas equivalent to the entire surface area of Earth. Structured areas in space could initially exceed deep ocean areas by a factor of 300. Biospheric areas could exceed structural areas by a factor of 1,800 if there were places to contain them. Of course, all the excess chemical elements in the life category are also useful in the plastics and structural categories. It seems reasonable to expect that many innovative architectural approaches will be possible with the 1.9×10^{24} tons of "plastics," 7×10^{24} tons of other materials, and creative uses of gravity, tension, and other means known in physics and engineering. Assuming 8×10^{16} people, these worlds might be sparsely populated compared with Earth today. If shallow oceans (one-tenth the depth of oceans on Earth) are reasonable, then there would be only 2 billion people/ocean world. Over time, controlled nucleosynthesis of the hydrogen and helium in the husbanded star could produce other desired distributions of heavier elements. Eventually, a billion or so worlds might contain on average only a few million people each. Artificial habitats rivaling in area and variety Larry Niven's "Ringworld" might be possible about most previously ordinary stars.

At the finish of phase II, assuming the organizing agencies required conditions suitable to humans, the s'homes might move in relatively close to the dwarf star and intercept most of its radiation. The life-filled s'homes might orbit close about the central star and radiate as a small composite sphere at a few hundred degrees Kelvin. Such spheres, either composite or solid, are referred to as Dyson spheres. It might be possible to detect the infrared emissions of solid-looking Dyson spheres about very large stars owing to the large total power and large distinct area that would be radiating waste heat. However, another model for the dispersal of s'homes about a husbanded star seems as likely and the effects harder to detect.

Power from the husbanded star might be intercepted by high efficiency power stations near the star. The stations could beam 80 to 90 percent of the energy to safer distances. Secondary swarms of facilities just outward from the power stations might use waste heat of the power stations for biological and other purposes. Net waste heat from the husbanded star and its surrounding machines would be 10 to 20 percent of its total power output and radiated from a tiny spherical region only a few times the size of the husbanded star. Individual and dispersed subswarms of s'homes could be located in orbits at great distances from the husbanded star. The s'homes would be dispersed to allow effective rejection of waste heat to cold space.

In phase III (fig. 4.3) the civilization would shift to the efficient use of power at approximately 2×10^{13} GW. This value is arbitrary and is used simply because it is generally consistent with an energy-rich society, which we can now envision constructing in space. Hydrogen would be upgraded over vast periods at the rate of 6×10^9 tons/year by fusion burning in the husbanded star. As we noted earlier, hydrogen would be converted at less than the rate new asteroids are now trapped by our Solar System.

The husbanded star and its surrounding swarms of life-filled s'homes, industrial volumes, and cryogenic globs would be almost invisible from a distance. Detection of individual complexes would be far more difficult using even infrared telescopes than envisioned for locating Dyson spheres near a sunlike star. It is doubtful that present space experiments for infrared observations would be adequate. However, it might be possible to detect the infrared emissions of thousands of complexes at once. The gravitational effect of the organized mass on surrounding visible masses might be the primary gross effect.

Figure 4.7 summarizes a dilemma that is becoming increasingly important in astronomy and cosmology. The farther a planet is located from its central star, the slower it moves along its orbit (note v_1 and v_2). One would think the same thing happens for stars in the distant regions of a disk Galaxy as they move in their millions years paths around the galactic center or core. However, this is not what is observed (note (1) and (2) in fig. 4.7). Rather, in our own and neighboring galaxies the stars progressively farther from their respective galactic core appear to move around that core with roughly a constant velocity ($v_1 = v_2$). There must be a large quantity of mass that is of extremely low luminosity or even nonluminous distributed approximately spherically around the center of the Galaxy and extending out past most of the luminous stars in the disk. Bahcall notes that limits on the accuracy of light curves obviate some of the interpretations of galactic mass distributions.

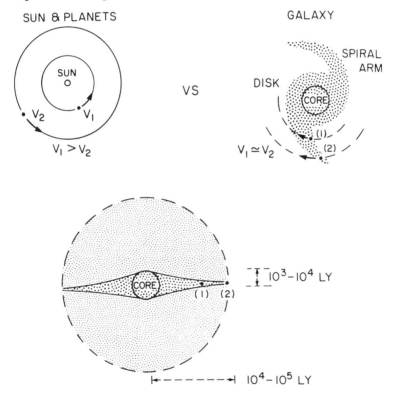

Fig. 4.7. The farther a planet is from the sun, the slower its orbital speed. Stars orbit around the core of their galaxy at the same speed regardless of their distance from the core. The cause of this must be that galaxies are surrounded by a halo of invisible but gravitationally active mass.

The form taken by the nonluminous mass is not understood. It is not gas or dust. Differential mass distributions of stars in the disk of the Milky Way Galaxy and in globular clusters are best known for stars between about 1 and 60 solar masses ($M_\odot$). It is not possible to reliably extrapolate available data to the white dwarf ($0.08\ M_\odot$ range or smaller). However, the differential distributions do appear to flatten out in the low-mass range which would preclude most of the stellar masses being composed of white or brown dwarfs. Theoretical models by Scalo for the evolution of gas clouds toward prestellar collapse indicate a paucity of stellar objects less than about $0.5\ M_\odot$ in size. Canizares argues that the dark halo matter is not composed of 0.01 to $100,000\ M_\odot$ objects. If such nonluminous objects were abundant in the halos of galaxies, then they would act as gravitational lenses. Many more distant galaxies would appear as optical doubles than are actually observed. Several other exotic types of mass have been suggested, including heavy neutrinos, gravitations, and so on.

An intriguing possibility, in view of the foregoing discussion of industrialization, is that the nonluminous, or "missing" mass in the halos of disk and some elliptic galaxies is composed of husbanded stars, the associated industrial facilities, s'homes, and stores of hydrogen, helium, and other elements. Very high precision astrometry could be directed to luminous stars in the halo of our Galaxy to watch for deflections corresponding to otherwise invisible masses of dwarf size. Extremely high resolution and highly sensitive telescopes might detect complex occultations of more distant photon sources by the sculpted and condensed masses. It may be that some of the industrial stars are converted into various types of gravitational machines. If so, there may be extremely distinctive emissions that can be detected and lend high confidence to the presence of other civilizations in the universe. Perhaps some of the compact pulsars or even superluminal sources are, as was first speculated, evidence of artificial activity on less than the scale of a Solar System. Gale and Edwards and others have argued that if advanced civilizations are present and active in the Galaxy or universe at large, it may be difficult over a short period of observation to distinguish their activities from natural processes.

Phase IV, the end of hydrogen burning, is not the final phase of star lifting. Lifetimes approaching hundreds of million billion years might be possible about Sol-type stars (see line 5 in fig. 4.4 and the area labeled "adjustment region"). It may be possible to extract power by dumping mass into black holes. Civilizations loosely organized over a Galaxy might even bring into being the extreme conditions thought to exist at the center of galaxies for recycling portions of the energy and matter. Frautschi has argued that entropy can be increased over vast realms of the universe by cooperative agencies that carefully dump matter into black holes. Such supergalactic empires would have to cooperate over far longer times than local civilizations husbanding dwarf stars or small black holes.

The concerns of some future supercivilization (in parenthesis) might parallel those of the present day:

> Today (10,000,001,985 A.D.) the *Los Angeles Times (Galactic Era)* reported on problems of dumping runoff (ultrahigh velocity ejecta) from the heavy snowfalls (recycled mass inputs) of last winter (the last galactic cycle). It was reported that some farmers (new civilizations) in the San Joaquin Valley (galactic disk) were resentful of the floods (shock waves) that were clearly going to occur. Strident voices will certainly be demanding that higher regard by the legislature (galactic council) be given to the longer-term needs for the safety and stability of those living in the low regions.

As Hewitt has suggested, perhaps the steady state-model of the universe should be reexamined. In an expanding steady-state universe, matter would be created at the rate of 10^{-44} kg/m^3/sec. This assumes a Hubble constant of 75 km/sec/megaparsec. If this matter could be collected and converted to energy, it would provide 8×10^{15} GW/light-year^3 of power. Assume there are 1×10^{12} solar masses of nonluminous matter within a sphere 100,000 light-years in radius about galaxies like the Milky Way or Andromeda. This indicates a spatial density of 1 solar mass/4,000 cubic light-year^3. There is about one visible star per cubic light-year in our neighborhood of the Milky Way Galaxy. If an advanced civilization were associated with a given solar mass and could accumulate the new mass created within its volume it would reap about 3×10^{12} tons/year or produce 3×10^{20} GW indefinitely.

Interstellar Migrations

How would interstellar migration proceed? Will there be, or has there been a headlong plunge from one star to many? Our first impulse is to envision beings moving to new and unknown stars and starting anew in a manner analogous to humans voyaging from one island to another in a vast ocean (see the papers by Finney and Jones in this volume). The process might be diffusionlike with many starts, stops, and even reversals. Newman and Sagan have argued that diffusion may be very slow owing to the large number of natural planets. The distractions to direct migration may be far more severe than they estimate. We argue that the stars themselves may be the primary targets—not the trivially small particles (planets, asteroids, comets, and so on) associated with the stars. A curious dilemma presents itself. If star lifting can occur, then a civilization can assemble extremely long-lived communities with enormous (compared with Earth) resources. A husbanded star could be surrounded by exquisitely intricate and extensive sets of s'homes and other constructs to beguile and distract even the longest-lived members of its various species. Husbanded regions and the universe itself might be filled with separate locales reasonably depicted by the cover of the *New Yorker* magazine—"The view from 42nd Street."

Most of the inhabitants may never see the natural stars or simply view them as primitive elemental resources too distant to be of primary concern. "Why go there?" they ask. "We've got it all here." Macrogovernments much longer lived than any of the governed may manage by distraction. They could open new local territories that had been closed for a few million years. It may be that space around a husbanded star is best thought of as a Meninger Sponge (fig. 4.8), which is composed of almost an infinite number of receding and self-similar recesses that interest and deflect the attention of its numerous civilizations inward as suggested by Mandelbrot. Which is most likely to migrate—beings or knowledge?

Travel at relativistic velocities between civilized, cooperative centers would be possible and even relatively cheap if energy-conservative devices for ejection and braking of transport craft were established. In principle this could be done with enormous electromagnetic mass drivers or gravity machines composed of sets of black holes or neutron stars rotating around each other in complex patterns. Enormous energies and peak power levels are necessary to sustain interstellar connections. The basic devices and systems of intercourse would have to be highly efficient and carefully directed. The situation is analogous to high-tension power lines on Earth, which convey the power equivalent of a raging forest fire quietly and steadily through forests and past homes. Cooperative and unobtrusive systems must generally be used. Only when a system breaks down, is interfered with, or is used to initiate a new extension would the effects of directed stellar energies be likely to be noticed at a distance (perhaps quasars or superluminous sources?). There is a price to be

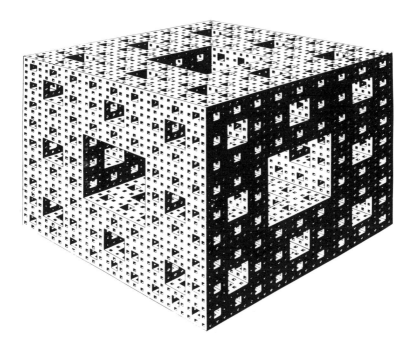

Fig. 4.8. Constructed habitats about a husbanded star may provide enormously re-entrant surfaces reminiscent of the fractil volumes of a Meninger Sponge like the one illustrated here.

paid for extremely high-volume, two-way transportation that uses energy-conservative systems. Momentum must be conserved. The two stations will move apart. If transport is extremely widespread, then spreading will be widespread also. Transportation can in principle spread apart the inhabited locations in the universe. Could widespread relativistic travel be relevant to the expansion of the universe or the form of extremely old spherical clusters, which tend to be clear of dust and gas?

"Hard-wired" interstellar railroads might be a way to provide cultural mixing within a sector of the Galaxy several tens to thousands of light-years across where reasonably quick two-way conversation is impossible. Civilizations could send many of their entities every year to husbanded systems specially equipped to accommodate emigrants from millions of worlds. Within that Solar System social intercourse could proceed rapidly without years-long delays. Later a few of the original emigrants, but primarily their progeny, could travel back to their original s'homes to complete the cycle. This is a prolonged type of the Grand Tour pursued by young British gentlemen in the eighteenth and nineteenth centuries. A hierarchical system could maintain slow mixing even over the scale of a Galaxy.

A curious thought presents itself. Is there an extremely deep motivation for intelligent life forms to spend most of their lives traveling near the speed of light? Advanced and ancient civilizations might provide this basic right to their citizens. If so, there might be enormous reluctance by the near-c inhabitants to visit Earth or similarly unequipped worlds that live in the slow lane.

One-way interstellar migration by isolated craft may be of little interest or utility. However, entire husbanded systems might profit and grow if they happened to be on trajectories that would take them slowly through dense clouds of dust and gas, nebulae, or even near massive giant stars with strong stellar winds. The husbanded system could gain immense materials resources. Would it be necessary to obtain mining rights from the galactic council, or are there rules of the road that apply? On a longer time scale there are probably stars that are randomly ejected toward other galaxies. Star-lifting civilizations could attach themselves to such stars and slowly diffuse outward. The lifetime of the Galaxy and such questions as the time to gravitational closure of the universe become relevant.

Figure 4.9 presents one aspect of first contact between two previously isolated cultures. Note the first two frames (t and t'). If the two cultures reach the contact point by means of relativistic probes and the distance R' as measured in light years is large, then the technologies and beings at the contact point will be relatively ancient (obsolete) by the standards of the central regions of each civilization. On a larger scale, say, intergalactic, this effect becomes more significant.

Omega, Myths, and the Long Rangers

Stellar husbandry might be related to the vast cosmological question of whether the universe is closed or open against gravitational expansion and possibly to whether the universe originated in a big bang or has always existed. Debate is vigorous on the basic issues. The ratio of accounted-for mass in the universe to the mass required to overcome expansion of the universe by gravitational attraction is termed *omega*. Omega greater than one ensures closure and possibly a cyclic universe. Omega less than one indicates a universe that expands forever and possibly gets colder. Omega exactly equal to one is probably a subject amenable to great debates. Nonluminous or

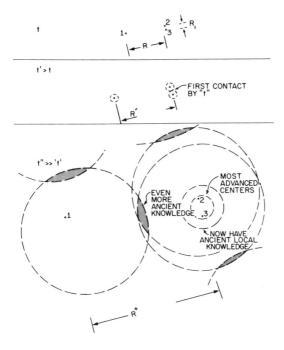

Fig. 4.9. Ancient knowledge will be the primary intelligence communicated between civilizations in an expanding universe.

missing mass in galaxies is relevant to expansion of the universe. The luminous stars and gas clouds and inferred interstellar gas (which is nonluminous but affects observable starlight) are estimated to provide approximately 10 percent of the mass necessary for universe closure. If organizing entities have converted a large fraction of the stellar mass in our own and other galaxies to husbanded stars and surrounding structures that use energy in a highly efficient manner, then omega near or in excess of one (locally or everywhere) may be the case. Discovery and confirmation of ancient myths may be one approach to evaluating the age of the universe and the extent of the presence of other civilizations. How might civilizations communicate over intergalactic scales of time and space? Figure 4.10 suggests a generic approach.

Pairs of large s'homes or highly intelligent automatic probes are boosted as close to the speed of light as possible. The pairs (or Long Rangers) are directed toward other galaxies. On board the pairs, which can communicate with one another, time goes by very slowly.

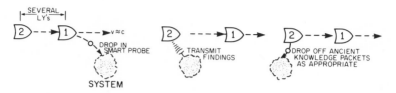

Fig. 4.10. Pairs of relativistic vehicles could spread and acquire knowledge across small portions of the universe.

Outside the pair the universe accumulates experiences rapidly. The pairs are repositories of their original knowledge and whatever they acquire en route. As the lead craft approaches an interesting system, it ejects a small probe (the craft must decelerate the probe to a small fraction of c; it actually fires it backward). The probe enters the system and automatically explores activities therein. Its findings are radioed to the second craft in sufficient shiptime to permit on-board evaluation of the findings. As appropriate, the second craft can drop off additional packets or send instructions to the probe to perform other tasks. One such task might deal with the construction over a long period of time of transportation and communication facilities to eventually support cooperative and energy efficient interstellar travel. Might such a facility be present in our Solar System? I have long been intrigued by the projective engineering book, *The Spaceships of Ezekiel*, written by Josef Blumrich, a former head of propulsion engineering at the Marshall Space Flight Center.

We note in passing a comment made by J. van Allen in the mid-1960s. As a craft approaches the speed of light, even the slow-moving interstellar gas appears to be of cosmic-ray energies. Thus, the traveling pair would be subjected to intense radiation and could be expected to emit large fluxes of radiation as they plow through the interstellar media. Might the pairs accumulate a portion of the interstellar medium and fusion burn it to overcome drag and provide on-board power as first suggested by Bussard in 1960? Is this relevant to quasars?

Suppose the universe is far older than we suppose and is expanding; that is, we live in a steady-state universe. This means that two stars separated by 100 million light-years 100 million years ago would be much farther apart now. Refer to the third frame of figure 4.9. Hubble's constant (Hb) is about 100 km/sec/million parsecs or 0.0001 c/Mly (Mly = million light-years), where c is the speed of light. Thus, the two stars would be receding from each other at 0.03 c and during 100 million years would change from 100 Mly to approximately 130 Mly apart. Over a longer period of time, say 1 billion years, the increase in velocity would have to be accounted for. The initial separation velocity would be 0.03 c, and the later velocity would be 0.3 c for an average of 0.16 c. Over the billion years the separation would increase from 100 Mly to 160 Mly. In the last two frames of figure 4.9 we see that over very long periods of time the first elements of the various physically isolated cultures to come into contact will be the most ancient elements, which have traveled outward at relativistic velocity. There will be types of knowledge and species inversions as explored by Ursula LeGuin. At a given advanced center, current knowledge and the most ancient knowledge may both be known with some accuracy. The events and knowledge in the middle histories will be less interactive. If the universe is not closed, then we might have access to the earliest experiences of the first races to conduct interstellar migrations.

Scared to Look?

Might it be that large-scale interstellar migrations, if they exist, are highly efficient, organized, and very difficult to detect? Cooperative systems would be required for efficiency. Extensive cooperation between complex systems, such as post-star-lifting civilizations, may be the natural results of tit-for-tat learning incorporated by most subsystems at a very early stage, as suggested by Axelrod and Hamilton. Advanced visitors to more primitive solar systems (prestar lifters) might be strongly inclined to maintain a relatively low profile. It might be advantageous to wait until the complex entities (for example, terrestrial societies) displayed potentials for appropriate responses of sufficient complexity before making extended contacts.

As Arthur C. Clarke has suggested, if we have been visited in the past, then it is likely that a terminus or the devices to build such a terminus are within the Solar System. Sporadic, unsupported visits would be extremely expensive. I suppose we can presume ourselves to be off any main lines. Visitors to Earth might know in advance that they would face enormous privations we presently have not even imagined.

Very advanced and long-lived civilizations would likely be extremely efficient in use of their primary stellar materials and energy. In the steady state they would be difficult to observe and would likely be busy with their own affairs. Collective gravitational effects, thermal emissions, and increasing obscuration of distant sources by dark halo masses might be the most directly detectable effects. It might be very easy to mistake directed activities in the large scale for completely natural processes. Civilizations making the transition from planet-bound to stellar husbandry might provide the most visible evidence of individual civilizations. It seems likely that their activities could be mistaken for natural but puzzling behavior by strange types of stars. If the universe is very ancient (not closed and cyclic), then relativistic travelers representative of the most ancient technical achievements could be journeying through the universe spreading knowledge of the ancient times. If the universe is cyclic we may be the first in this neighborhood, this time around.

The effect of habitations in a densely but quietly populated Galaxy or universe might be detectable. We might learn many lessons relevant to our own situation as our civilization approaches the possibility of permanent occupancy of space. Our searching for other life beyond Earth may be analogous to a needle searching for a warm haystack.

Summary

Interstellar migrations will require control of large quantities of energy and flows of power. In extreme cases the power could rival the output of a star. Large and long-lived technological societies may find it easier to undertake interstellar migration than will small, less-advanced civilizations. Civilizations that have major activities throughout their solar system(s) will be more likely to support interstellar journeys than civilizations bound to a single planet. Following is an exploration of how industrial societies on Earth might expand into the solar system, convert the large natural resources of solar system mass and energy to human control, and support interstellar travel. This model inspires exploration of cosmologies in which life is intertwined with the evolution of the universe.

Human industry now processes mass (primarily carbon fuels, sand and gravel, as well as processing and life-support chemicals) on the scale of planetary wide photosynthesis and geochemical concentration of economic elements. Human industry is still growing. So many people now live on Earth that artificial (i.e., industrial) means are absolutely required to sustain them. Barring catastrophe we will be driven to space to efficiently obtain solar energy for our growing, planetary-wide, artificially supported societies. A confluence of increasing abilities and efficiencies loom ahead. Solar energy will not require processing of mass as in the burning of carbon fuels. Rather, hydrogen burning in the sun will provide the power. Our energy industries will become far more efficient and supply far cheaper energy throughout most of the Solar System. Smart machines will increasingly expand the control of people over dumb mass and natural energy. Applicable skills will grow explosively on and off Earth. In space large quantities of mass will be converted to life-supportive devices by macrofactories. Limits to human growth will be effectively removed for centuries as populations grow in space in engineered habitats (s'homes), which can be far richer than Earth. In 500 to 1,000 years the cold mass of the Solar System could be converted to human control, humanity would be spread throughout the Solar system, planets would be minor considerations to the macrosocieties, and it could be necessary to modify the sun for more efficient utilization.

A star such as Sol wastes most of its energy to dark space. Sol is also burning itself out relatively rapidly. The smallest stars, white dwarfs, burn 2,000 times slower than does Sol. A basic option for an advanced human civilization is to take apart the sun a layer at a time. This could be done by large scale magnetohydrodynamic machines powered by 10 percent or more of the captured solar energy. These machines could be made of materials from the planet Mercury using self-reproducing (i.e., exponentially growing) machine systems. The hydrogen and helium would be saved and reintroduced into the new Sol-white dwarf as needed. Heavier elements could be initially converted into millions of oceans, trillions of areas equivalent to Earth in size, and more than 2×10^{21} people. Nucleosynthesis of primordial hydrogen and helium could produce any desired ratios of life to water or construction materials. Conversion of Sol (star lifting) would require approximately 300 million years. It could start as soon as 150 years from now, assuming a 20 percent annual growth rate of materials handling in space.

A civilization burning 1 Sol mass in a white dwarf could live 1,000 times the presently estimated age of the universe and provide 8×10^{16} people each with 200 kw of power and more than 30 million tons of support mass (excluding the hydrogen and helium). Establishing black holes for complete conversion of mass to energy would extend civilization lifetime to 2 million times the present age of the universe. Advanced civilizations such as these would be very difficult to detect from a distance.

Astronomical observations have recently revealed that many galaxies are surrounded by a halo of invisible mass, the constituents of which must be smaller and far darker than white dwarf stars and which cannot be free gas or dust. This nonluminous mass is approximately ten times greater than the luminous star, dust, and gas masses in the disks of the respective galaxies. Is it possible that the nonluminous mass could be the habitats, converted and concealed white dwarfs, and macromachines of advanced civilizations? Large astronomical observatories in space could clearly resolve such civilizations. Nonluminous mass throws into complete doubt the observational underpinnings of big bang cosmology. The hydrogen-to-helium ratio is no longer known. It may be appropriate to reinterpret the 3° black body radiation in the universe as not resulting from the big bang. It seems appropriate to reconsider the continuous creation hypothesis of cosmology. Olber's paradox will have to be reexamined. Opacity of civilized structures may be significant on the cosmological scale. If continuous creation of matter does occur, then advanced civilizations in the halo of galaxies could obtain on average more than a billion tons of new matter (equivalent to 3×10^{20} GW of power) forever. New civilizations would be continually spawned as the universe expands.

Such advanced and long-lived civilizations could support energy- and momentum-conservative interstellar travel between cooperative centers. Such travel in principle could be relatively cheap. Perhaps large fractions of the living population of the universe permanently exist in structures moving near the speed of light and enjoy some benefit of time dilation, which is not now appreciated by our primitive and planet-bound intellects. Perhaps very ancient knowledge of now extremely distant realms of the universe can become available to humanity if such long-lived civilizations exist. The chances of detecting and learning from advanced and energy-efficient civilizations are much better using observatories remote from Earth in the darkness of deep space. Such observatories are completely feasible to the technologically advanced nations of Earth. Earth's advanced civilizations could profitably direct some of their energies outward to search and possibly learn.

ENJOY!

ACKNOWLEDGMENTS

It is a pleasure to acknowledge the support of the California Space Institute and the Los Alamos National Laboratory of the University of California. My special thanks go to Paula for unflinching help and long-term patience.

References

Allen, C. W. 1963. *Astrophysical Quantities.* W. Clowes, London. Pp. 291.

Axelrod, R., and W. D. Hamilton. 1981. The Evolution of Cooperation. *Science* 211 (March 17): 1390-1396.

Bahcall, J. N. 1983. The Ratio of the Unseen Halo Mass to the Luminous Disk Mass in NOC 891. *Astrophysical J.* 267 (Apr. 1): 52-61.

Blitz, L., M. Fich, and S. Kulkarni. 1983. The New Milky Way. *Science* 220, 4603 (June 17): 1233-1240.

Blumenthal, G. R., H. Pagels, and J. R. Primack. 1982. Galaxy Formation by Dissipationless Particles Heavier Than Neutrinos. *Nature* 299 (Sept. 2): 37-38.

Blumrich, Josef F. 1974. *The Space Ship of Ezekiel*. Bantam Books, New York.

Bussard R. 1960. Galactic Matter and Interstellar Flight. *Astronautica Acta.* 6: 179-194.

Canizares, C. R. 1982. Manifestations of a Cosmological Density of Compact Objects in Quasar Light. *Astrophysical J.* 263 (Dec. 15): 508-517.

Clayton, D. R. 1968. *Principals of Stellar Evolution and Nucleosynthesis*. McGraw-Hill, New York.

Clarke, A. C. 1982. *2010: Odyssey Two*. Del Rey/Ballantine Books, New York.

Criswell, D. R. In press, Cis-lunar Industrialization and Higher Human Options, *Space Solar Power Reviews*, (Preprint available, A-021, UCSD, La Jolla, CA 92093).

Criswell, D. R., and R. D. Waldron. 1982. *Lunar Utilization in Space Industrialization*, vol. 2. Edited by B. J. O'Leary, CRC Press, Boca Raton, Florida. Pp. 1-53, 226.

Davis, M., M. J. Geller, and J. Huchra. 1978. The Local Mean Mass Density of the Universe: New Methods for Studying Galaxy Clustering. *Astrophysical J.* 221 (Apr. 1): 1-18.

Davis, M., J. Tonry, J. Huchra, and D. W. Latham. 1980. On the Virgo Supercluster and the Mean Mass Density of the Universe. *Astrophysical J.* 238 (June 15): L113-L116.

Dicus, D. A., J. R. Letaw, D. C. Teplitz, and V. L. Teplitz. 1983. The Future of the Universe. *Scientific American* 248 (Mar.): 90-101.

Dyson, F. J. 1966. The Search for Extraterrestrial Technology. In *Perspectives in Modern Physics*. R. E. Marshak, .ed., John Wiley, New York. Pp. 641-656.

Dyson, F. J. 1979. Time Without End: Physics and Biology in an Open Universe. *Reviews of Modern Physics* 51, 3 (July): 447-460.

Faber, S. M., and D. N. C. Lin. 1983. Is There Nonluminous Matter in Dwarf Spheroidal Galaxies? *Astrophysical J.* 266 (Mar.1): L17-L10.

Fabricant, D., and P. Gorenstein. 1983. Further Evidence for M87's Massive, Dark Halo. *Astrophysical J.* 267 (Apr.15): 535-546.

Frautschi, S. 1982. Entropy in an Expanding Universe. *Science* 217, 4560: 593-599.

Gale, W. A., and G. Edwards. 1979. Models of Long Range Growth. In W. A. Gale ed., *Life in the Universe—the Ultimate Limits to Growth.* AAAS Selected Symposium No. 31. Westview Press, Boulder, Colo. Pp. 71-106.

Glaser, P. E. 1977. Solar Power from Satellites. *Physics Today* 30-69, and letters (July): 6-15, 66-69.

Goeller, H. E., and A. M. Weinberg. 1978. The Age of Substitutability, *Science* 191: 683-689.

Hunter, M. W., and D. W. Fellenz. 1970. The Hypersonic Transport—The Technology and the Potential. AIAA No. 70-1218, American Institute of Aeronautics and Astronautics, New York.

Jones, E. M., and B. R. Finney. 1982. Interstellar Nomads, LA-UR-82-3630, rev., Los Alamos National Laboratory, Los Alamos, NM 87545.

Le Guin, U. 1975. *Rocannon's World.* Ace Books, New York.

Mandelbrot, B. B. 1978. *The Fractal Geometry of Nature.* W. H. Freeman, San Francisco.

Melott, A. L. 1982. The Formation of Galactic Halos in the Neutrino-Adiabatic Theory. *Nature* 296 (Apr.): 721-723.

NASA SP-413. 1972, *Space Settlements—A Design Study.* U.S. Government Printing Office, Washington, D.C.

NASA SP-428. 1979, *Space Resources and Space Settlements.* U. S. Government Printing Office, Washington, D.C.

NASA CP-2255. 1982, *Advanced Automation for Space Missions.* Edited by R. A. Freitas and W. P. Gilbreath. National Technical Information Service, Springfield, VA.

Newman, W. J., and C. Sagan. 1981. Galactic Civilizations: Population Dynamics and Interstellar Diffusion. *Icarus* 56: 293-327.

Niven, L. 1970. *Ringworld.* Ballantine Books, New York.

O'Neill, G. K. 1974. A Lagrangian Community. *Nature* 250 (Aug. 23): 636.

O'Neill, G. K. 1975. Space Colonies and Energy Supply to the Earth. *Science* 190, 4218 (Dec. 5): 943-947.

Physics Today, 1983. New Inflationary Universe: An Alternative to Big Bang? 36 (June): 17-19.

Reynolds, R. T., J. C. Tarter, and R. G. Walker. 1980. A Proposed Search of the Solar Neighborhood for Substellar Objects. *Icarus* 44: 772-779.

Rubin, V. C. 1983. The Rotation of Spiral Galaxies. *Science* 220, 4604 (June 24), 1330-1344.

Sagan, C., and R. G. Walker. 1966. The Infrared Detectability of Dyson Civilizations. *Astrophysical J.* 144, 3: 1216-1218, notes.

Scalo, J. M. 1978. The Stellar Mass Spectrum. In T. Gehrels, ed., *Protostars and Planets.* University of Arizona Press, Tucson. Pp. 265-287.

Stapledon, W. O. 1930. *Last and First Men; A Story of the Near and Far Future.* Methuen, London.

Stapledon, W. O. 1937. *Star Maker.* (Both Stapledon volumes are available as a combined volume from Dover, New York.)

Strom, K. M., and S. E. Strom. 1982. Galactic Evolution: A Survey of Recent Progress. *Science* 216,4546 (May 7): 571-580.

Van den Bergh, S. 1981. Size and Age of the Universe. *Science* 213, 4510 (Aug. 21): 825-830.

Waldron, R. D., T. E. Erstfeld, and D. R. Criswell. 1979. The Role of Chemical Engineering in Space Manufacturing. *Chemical Engineering* 86 (Feb. 12):80-94.

Wetherill, G. W. 1979. Apollo Objects. *Scientific American* 240 (Mar.): 54-65.

Eric M. Jones and Ben R. Finney

FASTSHIPS AND NOMADS: TWO ROADS TO THE STARS

Clarke's Laws of Prophesy

First Law: When a distinguished but elderly scientist states that something is possible, he is almost certainly right. When he states that something is impossible, he is very probably wrong.

Second Law: The only way of discovering the limits of the possible is to venture a little way past them into the impossible.

Third Law: Any sufficiently advanced technology is indistinguishable from magic.

—Arthur C. Clarke, *Profiles of the Future*

Arthur C. Clarke is a notable science fiction author (*Childhood's End, 2001*), science writer (*The Exploration of Space*), and inventor, of the concept of communications satellites at geostationary orbit. His three laws of prophesy serve as excellent guides to those of us who dare speculate on the future of human technology, especially the rather grand vision of interstellar travel.

Stated another way, the laws of physics and chemistry as we know them permit many things. It is the job of creative engineers and scientists to apply natural law to practical problems. Natural law also prohibits certain things. For example, on Earth human beings are simply not capable of unaided flight. However, as Clarke suggests in his laws and in elaborations, our knowledge is limited. As we learn more about the way the universe works, apparently intractable problems may find eminently practical solutions. For example, modern technology has produced the strong, lightweight materials to make the ancient dream of human-powered flight a reality. .

Spaceflight is a reality. We have the technology in hand that will enable our descendants to populate the Solar System: chemical rockets, solar sails, and many other techniques. But what about interstellar travel? The distances to be crossed and the technical problems to be solved are much greater. There are those who echo the naysayers of previous generations: Interstellar flight, if not impossible, will always be immensely impractical.

Part of the problem is this: What do we mean by interstellar flight? Already we have sent four spacecraft on journeys out of the Solar System, using gravity assists from Jupiter and Saturn. Had we aimed any of the four (*Pioneers* 10 and 11 and the two *Voyagers*) toward the nearest stars, they would have arrived in a few tens of thousands of years. However, if by interstellar travel we mean a journey to the nearest stars in less than a human lifetime, the problem becomes much more difficult.

Since neither of us is "a distinguished but elderly scientist," we will elaborate our assertion that our descendants will eventually reach the stars. We offer two scenarios and expect that both will come to pass. One is a high road traveled by pioneers in the fastships of science fiction. These will be the journeys of no more than a few decades. The other is a slow meandering route among interstellar comets that will take thousands of generations to travel.

There is a relevant analogue from human experience. Suppose you want to get to Tierra del Fuego, the cold, storm-swept islands that form the tip of South America. One way to go is by sea, just as the early European explorers did as they traced Magellan's route to the Pacific. Although they did not have the advantage in Magellan's day of the modern navigational techniques Captain Cook would pioneer 200 years later, these were still purposeful, long-distance voyages. Yet when the Europeans arrived at Tierra del Fuego, they found the place already inhabited. We now know that humans first crossed into the Americas by the Bering land bridge some 40,000 years or more ago and that a slow, southward diffusion over many generations eventually brought descendants of those first Americans to the southernmost point of the New World. In one instance people came more or less directly to Tierra del Fuego (if only to round Cape Horn or pass through the Straits of Magellan on their way somewhere else); in the other, people reached the island as a consequence of the steady spread of hunters and gatherers across the face of the planet. In the case of interstellar travel we expect that it will be the direct, purposeful voyages that will first cross the interstellar deep. We are already reaching for stellar Tierra del Fuegos and Tahitis in our imagination. But even should fastship technology prove to be immensely impractical or should the will to develop it lag, humanity has a second chance, for we think that travelers on the slow road will eventually attain the stars.

Interstellar Fastships

The basic fact with which a potential interstellar traveler has to deal is that the stars are very far apart. In the solar neighborhood the average spacing is 3-4 light-years. Some stars are likely to be more hospitable than others. As in the Solar System the settlers will need not only light from the star to provide energy but also planets, moons, asteroids, and comets to furnish building materials for a growing population. Current theories of star formation suggest that only single stars rather like the Sun will have families of the needed smaller bodies. If we restrict our interest to such stars, the average separation grows to about seven light-years. A journey lasting a lifetime or less will require ships traveling at least a few percent of light speed. Chemical rockets like the Saturn V and other conventional launchers don't even come close to such speeds.

Is fastship travel possible at all? A number of careful thinkers suggest feasible solutions. The basic problem is energy. Chemical rockets are insufficient because of a basic characteristic of their design. The launch energy is supplied by on-board fuel (which also provides reaction mass); the need to carry fuel exacts a terrible price. During each moment of the launch, the fuel being burned has to accelerate not only the useful payload but also the fuel to be burned during the remainder of the launch; hence, the enormous size of chemical rockets. An obvious solution is to avoid carrying fuel at all but rather to supply the energy from an external source. All the attractive schemes for achieving interstellar speeds use this principle.

A particularly elegant scheme combines ideas contributed by Clarke and Freeman Dyson, a senior scholar at the Institute for Advanced Studies in Princeton. Clarke has noted that great amounts of solar energy could be tapped by placing giant power stations (Clarke stations) close to the Sun. The collected energy could then be beamed by laser or microwave to other places in the Solar System. The collectors (probably very thin mirrors) would be very large (Fig. 5.1). A 500-passenger interstellar ship massing 500,000 tons could be launched with the power supplied by a mirror 3700 km across stationed at a tenth of Earth's distance from the Sun.

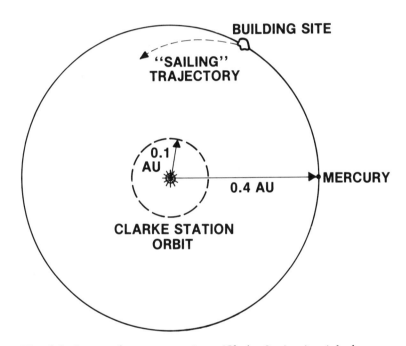

Fig. 5.1. Large solar-power stations (Clarke Stations) might be constructed from Mercurian materials launched from the surface of the planet by mass driver. The building site might be at one of the Sun-Mercury Lagrange points 60 degrees ahead or behind Mercury in its orbit about the Sun. The completed station could be sailed from the building site to the duty station at one-tenth the Earth-Sun distance (0.1 AU).

Dyson has recently pointed out a very efficient scheme for using this energy to launch an interstellar vessel—microwave sails. A microwave sail is a close cousin of the solar sails now being developed by the World Space Foundation and others. A solar sail works because it reflects sunlight (or a laser beam) and is accelerated in the opposite direction. If we take advantage of Newton's laws, a solar sail can even approach the Sun. What we need to do is have the leading edge (defined by the direction of motion around the Sun) farther from the Sun than the trailing edge (fig. 5.2). The result is a component of the acceleration opposed to the orbital motion; the sail then spirals (very slowly) inward toward the Sun. In the long term, solar sails look very attractive for long voyages within the inner Solar System, for example, for freight traffic. However, they are very impractical for interstellar voyages . Sunlight is too weak and even the thinnest sails imagined are too massive to produce the desired accelerations. True, laser beams could be used to concentrate the photon stream and increase the acceleration. G. Marx of Roland Ëtvö University, Budapest was the first to suggest laser-launched light sails, but laser-propelled sails are basically too massive for interstellar vessels.

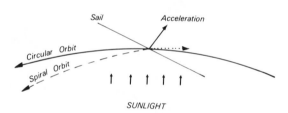

SOLAR SAILING

TOWARD THE SUN

SUNLIGHT

Fig. 5.2. Sunlight falling on a solar sail gives an acceleration perpendicular to the surface of the sail. If the leading edge (to the left for this counter-clockwise orbit) is farther from the sun than the trailing edge, there is a component of the accelera-tion (dotted line) opposite the direction of mo-tion. The sail then slows and begins a slow spiral toward the Sun. Such a scheme could be used to move a completed Clarke station.

Dyson has shown how to reduce the sail mass (actually, how to reduce the surface density). Use microwaves rather than sunlight or laser light! A good mirror has to be smooth on a scale of the wavelength of the photons to be reflected. Because the photons that constitute visible light, be it sunlight or a laser beam, have very short wavelengths indeed, light sails or telescope mirrors must have a surface of tightly spaced atoms. However, the photons that make up a microwave beam have wavelengths much longer than visible light, wavelengths of roughly a few centimeters, so the reflecting elements can be very widely spaced. A loose wire mesh would do nicely. The result is that the surface density of a microwave sail could be hundreds of times less than even the best light sails. Without going into detail, a microwave sail massing about 200 metric tons would be sufficient to send a party of 500 on their way to the stars. The sail would be enormous, nearly 6,000 km across, but gossamer thin. The sail mass would be only a few percent of the payload.

The great efficiency of microwave sails suggests that interstellar vessels could be rather substantial affairs, large enough to provide considerable elbow room and privacy for the settlers during the decades-long voyage and capable of carrying all the interstellar party would need to establish a self-sufficient community at the destination. To be a little more specific, we can take guidance from current designs of orbital space colonies, which suggest that each emigrant would need about 1000 metric tons, most of it cosmic-ray shielding. The most precious part of the cargo would probably be very capable (intelligent?) computers, which would supplement and greatly enhance the skill base of the human party. The ship would carry as well seed stock and other biological raw materials on which the success of the interstellar community will depend. Elsewhere in this volume William Hodges questions whether an interstellar expedition could carry enough skilled people and enough tools and seed stock to avoid the kinds of disasters that nearly befell England's Botany Bay colony in Australia or some of the early American colonies. We suspect that the fruits of the information and genetic revolutions will greatly ease the problems he raises.

The interstellar vessel could be very roomy indeed; probably rather like the space colonies proposed by Gerard O'Neill and his colleagues. The vessels might well include parkland interspersed with farms and clusters of residences. No submarine claustrophobia aboard the *Freeman Dyson*!

One potentially serious problem with our estimates is the fact that the 1000 tons/emigrant does not include any provision for slowing down at the destination. It just wouldn't do for the ship to whiz past the target star at a tenth lightspeed. Although Princeton scientist Cliff Singer has proposed an elegant method of sending fusion fuel ahead of the ship in a "pellet stream" (the fusion pellets being sent out of the Solar System before the ship itself is launched), such methods seem too prone to error to be entirely satisfactory. We suspect that inter-stellar emigrants would prefer some on-board system so that they could have more control over their destiny.

One class of solutions would be to take fusion engines and fuel to effect the deceleration, but such a scheme would extract the same type of mass penalty as conventional chemical rockets. In the best of circumstances the launch mass would increase by a factor of about 4. The launch energy would increase by the same factor. Although 4,000 tons/emigrant is not out of the question, there do exist other schemes that look more attractive. They include using the galactic magnetic field for breaking or using enormous scoops to gather fuel from the interstellar gas for a fusion ram jet.

The problem of deceleration at the destination should be solvable. We think it unlikely that a practical scheme will not be found in the next few centuries. Harkening back to Clarke's third law (advanced technology = magic), we recognize that exotic solutions that take advantage of unforeseen technological breakthroughs may make in-terstellar travel easy to achieve. The point we want to make is that contemporary knowledge suggests that there are ways of reaching the stars without invoking "magic." Dreams with plausible manifesta-tions tend to become reality.

The existence of plausible fastship concepts suggest that once the available technology base has grown sufficiently large, small bands of explorers and pioneers will make the leap between stellar oases. How large the movement of people might be depends, of course, on the cost. If fastship voyages require a significant fraction of the total human wealth, they will be few and far between. We can estimate the relative cost. The sun outputs enough energy to permit 50,000 emigrants to leave the Solar System each second (if that were the only use of the gathered sunlight)! If, by the time humanity is ready for the interstellar adventure, our descendants have managed to tap even a modest fraction of the solar output, they could easily afford emigra-tion at a rate sufficient to sustain the human expansion. If we take the figure of 500 men, women, and children, a number suggested by studies of breeding populations among surviving hunter and gatherer peoples, as the minimum size of a genetically and socially healthy population, and if we stretch the launch period over a year, a collector 3700 km across will suffice to launch one such party each year. The need to carry a deceleration mechanism would increase the cost, which could be partially alleviated by building the collector closer to the Sun. The reader may make a private assessment as to the feasibility of such a system; but keep Clarke's laws in mind and remember that human capabilities have a way of growing with time.

We are not invoking dramatic scientific breakthroughs, just engineering on a very large scale by a people, our descendants, already used to living and working in space.

We will not discuss here the settlement of other stellar oases in any detail. Those details will depend on the tricks our descendants will learn as they settle the Solar System. Suffice it to say that the scope will likely be as broad as the settlement of the Solar System, save only that habitable planets are not expected or necessary. That should be no hardship to a people whose ancestors had given up planet dwelling many generations before. If interstellar settlement happens at all, it will come after our descendants have learned to maintain self-sufficient communities detached from Earth's nurturing biosphere, learned to tap the knowledge and skill potential of advanced computers, learned to efficiently harness the energy that flows out of the sun, and even learned to extract useful energy from the fusion of atoms. We are on the verge of achieving all these things. With sufficient skill and patience we will attain the stars.

Just when we will be ready is somewhat difficult to predict. Careful attention to Clarke's Laws and the explosive growth of human capabilities in the last few centuries suggest that we could be closer to interstellar voyages than we are to the Age of Exploration. How close, we cannot say. Probably within the next few hundred years human settlements will be struggling to gain footholds in the oases nurtured by the nearest stars. Once established, will the stellar settlements spawn further steps in a migration that began a million years or more ago in East Africa? We think so.

For a time, whether the founding population of a stellar oasis is as small as a few hundred or is much larger, the settlers will be very busy getting themselves established, perhaps like the founders of West Polynesia who paused for more than a thousand years on mid-Pacific archipelagoes before moving on to expand throughout the Polynesian triangle. We expect that after a similar period the capabilities of an initial settlement will have grown to the point that fresh pioneering voyages could be undertaken and the oasis will have been sufficiently tamed that the urge to move outward will have arisen again. Jumps to stellar oases farther and farther from the Sun will create a diffusive expansion that could carry our descendants to virtually every suitable star in the Galaxy, rather like the Polynesian expansion that touched just about every speck of land in the vastness of the Pacific. Unless flagging motivation, competing life forms, or some other unforeseen circumstance bars the way, this process may well lead, millions of years hence, to a Galaxy full of human descendants.

Interstellar Nomads

But suppose all this is fantasy. Suppose that fastship travel is impractical even to the solar society our descendants might create in the next few centuries. Will humanity be forever bound to this one tiny corner of the Galaxy? We think not. There is another road to the stars, a road paved with comets, on which journeys between stars will take a few thousand generations. If there are no fastships, this slower human migration might fill the Galaxy in half a billion years.

By far the most numerous objects in the Sun's family are comets. The Sun probably has several trillion, most circling our star far beyond the orbit of Pluto, forming a great spherical cloud, the Oort cloud, named after the Dutch astronomer who deduced its existence in 1950. From time to time a relatively near encounter with another star perturbs the orbits of a few comets. Some are sent spinning off into interstellar space. Others are nudged toward the inner Solar System where, a million years later, chance encounters with the giant planets may send the comets out of the Solar System like the *Voyager* spacecraft. During the last century the Solar System has definitely lost several dozen comets and probably very many more. If this process is common in the Galaxy, interstellar space is liberally strewn with comets. The average spacing is probably of the order of ten times the Earth-Sun distance, about 10 astronomical units (AU), or 80 light-minutes. As Freeman Dyson has suggested, if interstellar comets are common, "the galaxy is a much friendlier place for interstellar travelers than most people imagine." We imagine then a human population living off cometary resources and diffusing slowly out into interstellar space.

Before we describe how this migration of a people we might call interstellar nomads would begin, let us describe the resources available from the comets and how the scarcity of those resources might shape the nomad communities.

A typical comet may not seem very hospitable, being an irregular chunk of ice and rock perhaps a kilometer or two across. In fact, we don't imagine that people will actually live on the comets but, rather, use them as a source of raw materials to build habitats and supply the other physical necessities of life. Except for hydrogen, which will be underrepresented, and the chemically inert noble gases like helium, which will be very scarce indeed, the naturally occurring elements will be present in roughly cosmic abundances, a vast supply of material resources. The greatest obstacle facing potential nomads is energy, which is very scarce in the interstellar deep. Nonrenewable sources would include deuterium to power fusion generators and the kinetic energy (energy of motion) of the comet, which could be extracted through interactions with the galactic magnetic field. Renewable energy sources would include starlight collected with gigantic mirrors and possibly cosmic rays (if anyone can figure out a practical scheme for catching them). Elsewhere we have estimated that the aluminum in a typical comet would be sufficient to build mirrors to collect a few hundred megawatts of starlight. Other, more abundant substances may prove to be more applicable for the mirror surfaces. We expect that power levels of 1 megawatt (MW) per person are reasonable and therefore a typical comet could support a few hundred people with starlight. A megawatt is about fifty times the current per capita use in the United States.

Energy is by far the scarcest resource in interstellar space; together with social and genetic factors it will determine the size and structure of the communities. There is energy to permit the growth of very large communities. Comets have random velocities of about 10-20 km/sec, mimicking the random velocities of the stars that eject them. In a sense the comets are the molecules of a gas stirred by random gravitational encounters with the stars embedded in that gas. Comets could be gathered into large clusters if velocity changes could be imparted, rather like maneuvering two spacecraft for a rendezvous in Earth-orbit. Only a few percent of the deuterium in the cometary ice would be needed to bring a new comet into a cluster. Such clustering would be necessary in the case of the relatively numerous small comets, which might not be able individually to support a self-sufficient community.

Clusters could not grow without bounds for the simple reason that the gathering process would soon deplete the stock of nearby fresh comets. Imagine, for example, that the islands of the Pacific float and that before the Polynesians arrived the islands were distributed at random. Suppose then that the Polynesians wanted to live in island clusters and could drag the islands together. As the groups grew the average spacing between them would grow as well. In principle, all the islands could be clustered at midocean but that would involve dragging a large number of islands long distances from the periphery of the ocean. A more efficient process (and one that would promote genetic and cultural diversity) would be to form a number of relatively small groups scattered over the ocean. The time and cost of gathering would decrease dramatically.

Just how large the interstellar comet clusters would grow will depend on subtle factors; we offer only crude estimates. The starlight mirrors would be very large; each 1 MW mirror would be about the size of the continental United States. Although the mirrors would be tended by autonomous maintenance robots, the nomads would have to live nearby in case something went wrong. After all, the mirrors and the energy they supply will be vital to the well-being of the community. Although we could imagine that the several hundred people who could be supported by the resources of a single comet might live in a single habitat, the mirrors supporting that community would be spread across about 150,000 km. Trouble with a mirror or robot on the periphery of the mirror array would mean a long trip, several hours at least. It would make more sense if the community were dispersed in smaller groups so that trouble could be reached in a shorter time. There are also social reasons for expecting the nomad communities to be divided into smaller coliving groups. For reasons perhaps stemming from our distant past, we seem to function best in groups of about a dozen adults, be it a hunter-gatherer band, a bridge club, an army platoon, a political cell, or any other of numerous examples. We imagine then that people would live in bands of about twenty-five men, women, and children, each band occupying a spacious habitat constructed from cometary materials. They would tend (with considerable help from robots and computers) a farm of starlight mirrors stretching across perhaps 30,000 km of space (fig. 5.3).

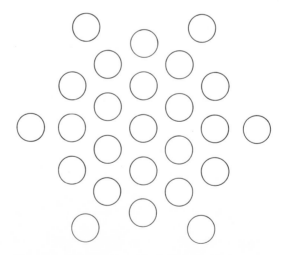

Fig. 5.3. This hexagonal array of starlight mirrors might support a co-living group of twenty-five adults and children. Each of the mirrors is about the size of the continental United States, and the entire mirror "farm" is about 30,000 kilometers across.

We model our minimal cometary unit on the oldest of human examples: the hunting and gathering band, the social organization within which our ancestors lived for millions of years. As first pointed out by Joseph Birdsell in his studies on Australian aborigines and illustrated by Richard Lee's sketch in this volume of the San people of the Kalahari, among surviving hunting and gathering people around the world these bands averaged roughly twenty-five men, women, and children. Within a band of twenty-five there are apparently enough men and women to effectively carry out the vital subsistence and childrearing functions, yet the band is not so large as to become unwieldy or fragmented when ranging over a hunting and gathering territory in search of scattered resources. However, although in everyday life these bands were essentially independent from one another, of necessity they formed into larger breeding communities averaging, on continent after continent, around 500 individuals. These tribal groupings, typically marked by dialect as well as gene-flow boundaries, with careful attention to marriage rules prescribing that mates must be sought among distantly related persons from outside the natal band, appear to be just large enough to avoid the genetic dangers of inbreeding and also to provide a seemingly necessary and enriching larger social environment. Thus, we expect that cometary bands would also cluster for purposes of healthy breeding and social enrichment in tribal groups of at least 500 citizens. We imagine clusters of some 20 cometary bands, the members of which would exchange marriageable youths and periodically meet to celebrate rites of passage, for calendrical observances, and for other social occasions.

Because of the great size of the energy-gathering mirrors, the members of such a cometary tribe, living within the constituent bands, would spread across some 200,000 km of space (fig. 5.4). Assuming power levels of about 1 MW per person, a tribe could easily live off the resources of a single typical comet or a few smaller singles (the smaller comets being more numerous) gathered into a cluster. The habitats and mirrors would 'fly in formation' with the comet cluster. Although the bands would live in separate habitats, they would gather at frequent intervals for tribal ritual. Should someone want to leave the tribe entirely, there would be opportunities every decade or so to transfer to a passing tribe. Such transfers would be rather expensive, because they would involve large velocity changes, but would not be beyond the energy means of the group. Such intertribal transfers might be fairly common if they serve to strengthen social ties between groups that might want or need to exchange

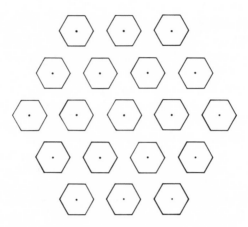

Fig. 5.4. Each of the hexagons in this figure represents one of the mirror farms shown in figure 5.3. The dots are schematic representations of the habitats occupied by each twenty-five-member "band." In reality the habitats would probably be much too small to be seen on this scale.

information or goods. Of course, the large relative velocities (10-20 km/sec = 2-4 AU/yr) between separate tribes would mean that they would be too far apart for physical exchanges after only a few years. However, close communications by electromagnetic means might persist for many generations.

This proven band-tribe structure is the smallest grouping that seems practical for any human community. Although, as we have just mentioned, information exchange would always be possible with a far larger number of people by sophisticated electromagnetic communications channels, larger communities may also develop. Perhaps the diversity of skills needed to support an interstellar community will require larger groups. However, estimates of the energy cost of travel across a sizable community suggest that the largest practical population is of the order of 100,000 people. A one-day trip across such a "city" would cost about a megawatt-yr. If, as we have assumed, power generation is of the order of 1 MW, frequent trips across such a city would strain the resources of the community. We therefore suggest that communities of interstellar nomads, living off the resources of clustered comets, will range in size from 500 to no more than 100,000.

How then might this comet-based human migration out of the Solar System proceed? We imagine that within a few hundred years exploitation of Solar System resources will have carried some adventuresome representatives of our species out to the Oort cloud. Because the Oort cloud comets are relatively close to the Sun, rather large communities could be supported by sunlight. O'Neill has estimated that the mass of solar collectors would have to equal the habitat mass at about 2,000 AU, well into the Oort cloud. There the interstellar comets may be sprinkled at a ratio of about one per several thousand solar comets. As the Oort cloud population grows, there will be a temptation to hitch a ride on an interstellar comet. The pioneers might well be a small group of young adults attracted by the adventure of it all. They would drift outward slowly, at about 2-4 AU per year. The group might grow rather quickly (as Birdsell suggests) and would soon be faced with the choice of group fissioning or gathering other comets. We suspect that both options would be exercised from time to time. Eventually, there would be an expanding sphere of nomads. At the frontier the gathering and fissioning options would both be possible, whereas closer to the Sun gathering and fissioning would end and some sort of steady state, a mix of large and small communities, would arise.

After about 50,000 years something dramatic may happen.

When the Vikings and, later, Columbus crossed the Atlantic, two branches of the human family that had been isolated for a similar period were reunited. It was not, of course, a trouble-free reunion. If, as we suspect, fastship travel is practical in some form, long before the interstellar nomads will have gotten very far out of the Solar System, fastship pioneers will have founded settlements in the nearest stellar oases. A thousand years later or so their descendants will have begun a cometary migration. About 50,000 years from now these two populations should meet roughly midway between the Sun and the nearest stars.

There are probably good science fiction tales to be told about these meetings. The cultural and genetic diversity implicit in a large number of small communities suggests that the parties will find one another strange. However, we do not expect any surprise encounters. Because the comets drift at only a tiny fraction of lightspeed, the two flows of nomads should be well aware of each other's progress. Communication between approaching communities would be possible long before physical contact. Perhaps our descendants will have learned something from the sometimes disastrous historical contacts that Crosby describes. We leave the implications to science fiction and to posterity.

We have described in general terms how human beings might make a living as nomads wandering in the interstellar deep. We have tried to be cautious about extrapolations from present technology and have stayed close to admittedly sketchy modern concepts of interstellar resources. Just as the Solar System has proven to be a very different, but still wonderful, place from what we had suspected just a few decades ago, the interstellar deep will likely be stranger than we imagine—but perhaps more marvelous too and more hospitable to humankind.

Colonization of the Galaxy

We began this essay with Clarke's three laws of prophesy. The Master cautions against timidity. We could have indulged in flights of fancy, say, the warp drive that speeds Captain Kirk's *Enterprise* from one adventure to the next. Perhaps such things are in the cards. Certainly, the universe has many surprises in store for us.

However, our immediate purpose has been to explore the possibility that the very ancient human migration that began so long ago in East Africa might take our descendants beyond the terrestrial cradle through the Solar System to the stars. Very advanced technology (Clarke's "magic") might reduce the exercise to trivialities. We have tried to show that interstellar travel by one or both of two basic methods is feasible. We think that modest extrapolations of current trends in technology—in genetics, computers, and energy use—lead to the conclusion that interstellar migrations are a plausible expectation. Journeys to the stars and the establishment of viable human communities beyond the borders of the Solar System require no magic—only time, skill, patience, and continuity with the past.

References

Birdsell, J. B. 1979. Ecological Influence on Australian Aboriginal Social Organization. In I. S. Bernstein and E. O. Smith, eds., *Primate Ecology and Human Origins*. Garland Press, New York. Pp. 59-68.

Clarke, A. C. 1977. *Profiles of the Future*. Fawcett Popular Library, New York.

Drexler, K. E. 1979. High Performance Solar Sails and Related Reflecting Devices. In *Fourth Princeton/AIAA Conference of Space Manufacturing Facilities*. American Institute of Aeronautics and Astronautics, New York. Pp. 431-437.

Dyson, F. 1979. *Disturbing the Universe*. Harper and Row, New York.

Dyson, F. 1982. Interstellar Propulsion System. In M. H. Hart and B. Zuckerman, eds., *Extraterrestrials: Where Are They?* Pergamon Press, New York. Pp. 41-45.

Finney, B. R., and E. M. Jones. 1983. From Africa to the Stars: The Evolution of the Exploring Animal. *Space Manufacturing 1983. Advances in the Astronautical Sciences* 53:85-104.

Forward, R. L. 1982. *Roundtrip Interstellar Travel Using Laser-Pushed Lightsails.* Hughes Research Laboratories Research Report no. 550.

Jones, E. M. 1981. Discrete Calculations of Interstellar Migration and Settlement. *Icarus* 46: 328-336.

Jones, E. M. 1982. Estimates of Expansion Timescales. In M. H. Hart and B. Zuckerman, eds., *Extraterrestrials: Where are They?* Pergamon Press, New York. Pp. 66-76.

Jones, E. M., and B. R. Finney. 1983. Interstellar Nomads. *Space Manufacturing 1983. Advances in the Astronautical Sciences* 53: 357-374.

Marx, G. 1966. Interstellar Vessels Propelled by a Terrestrial Laser Beam. *Nature* 201, 22-23.

Oort, J. H. 1950. The Structure of the Cloud of Comets Surrounding the Solar System and a Hypothesis Concerning Its Origin. *Bulletin Astron. Instit. Netherlands* 408:91-110.

Singer, C. 1982. Settlements in Space, and Interstellar Travel. In M. H. Hart and B. Zuckerman, eds., *Extraterrestrials: Where are They?* Pergamon Press, New York. Pp. 46-61.

Wilkening, L, ed. 1982. *Comets.* University of Arizona Press, Tucson.

SECTION II

Demography and Economics: Growth of the
Human Tribe

INTRODUCTION

Human expansion into space will involve much more than just an adroit mating of technology and resources. It will depend ultimately on the ability of fledgling communities in space to become self-sufficient. If space colonies forever remain dependent upon support from the homeland, expansion into and beyond the Solar System would hardly seem possible.

England's Botany Bay Colony, founded in 1788 as a dumping ground for the overflowing population of British prisons, provides a terrestrial example of how a colony may stagnate when tied too closely to the founding population but may flourish when a measure of self-sufficiency is achieved. Although the originators had argued that a self-sufficient prison colony would greatly reduce the financial burden of dealing with convicts, others in the British Establishment could not square this vision with their own conception of necessary punishment. After all, they argued, news of prosperous farmer-convicts would incite to crime those left behind in the burgeoning slums of England's cities. So His Majesty's government decided that convicts would receive no land grants. They were to subsist primarily on meager handouts from England.

Nonetheless, the Australian settlement did eventually succeed, although the outcome was in doubt for several decades. In time, practicality and personal greed won out. The introduction and expansion of sheep ranching between 1810 and 1820, then the discovery of gold thirty years later, contributed to success. Less obvious, but more important, were the decisions of freed convicts, nonconvict immigrants, and members of the New South Wales Corps to make their future in Australia.

During the early decades the European population of Australia grew at the phenomenal rate of 8 percent per year. About three-quarters of the increase was due to immigration (mostly involuntary), but the rest was natural increase, that is, children were being born in the new settlement. Men and women decided to become "Australians" and to put down roots. In many ways the growth of the Australian settlement paralleled the American experience and suggests some trends we might expect to see in the growth of space settlements: an early period dominated by immigrants struggling to gain a foothold, followed by a period of increasing self-sufficiency and economic viability in which natural increase of the population plays a greater role. However, critics might argue that these two examples both describe only the experience of European colonization, and just its Anglo-Saxon variety at that. Perhaps space colonies derived from various modern societies would grow in very different ways. Might not the trend of very small population growth seen today in the developed world carry over into space? Perhaps natural increase of the space-living population would be so slow that the colonies could never become self-sufficient?

In the following papers two students of the human tribe examine these and related questions.

In the first paper Joseph Birdsell, Professor Emeritus of Anthropology at UCLA, examines the growth of very small, genetically isolated communities. Using examples from his research among Australian aborigines and other peoples, Birdsell argues that with proper care (planned or otherwise) in selecting the founders, viable communities can spring from very small beginnings. The reader should look forward to the story of "Afghan," a man who surely holds a special place in the history of human reproduction. Birdsell's examples suggest that even in very difficult circumstances, humans are fully capable of doubling their numbers every generation (roughly 2.7 percent per year). These examples may have particular significance for the growth of the nomad populations we sketched in the last paper.

Although the examples Birdsell describes suggest the possibility of success for individual population experiments, we do not imagine that the settlement of space will involve a handful of experiments any more than did the human expansion out of Africa. Ultimately, the settlement of space will be a statistical process governed on the small scale by personal choices and on the large scale by average behavior of many people. In the second paper of this section, Kenneth Wachter discusses general trends in demographics (the growth and decline of populations) that might apply to space. In particular, he emphasizes the great difficulty encountered when we try to extrapolate short-term trends. How important are phenomena like the post-World War II "baby boom"? Do population increases in frontier circumstances continue trends in the homeland? How important are the successful few? What is the minimum size of a viable population?

In the final paper of this section William Hodges discusses some economic issues that tie in closely with these demographic questions. Although Hodges (a Ph.D. candidate in economics at Berkeley) did not attend the Los Alamos conference, his detailed comments on a draft of Wachter's paper raised issues that deserve attention. We have extracted the essential arguments from his longer submission and offer some comments following his paper.

References

Clark, C. M. H. 1950. *Selected Documents in Australian History: 1788-1850*. Angus and Robertson, London.

Shaw, A. G. L. 1972. *Story of Australia*, 4th ed. Faber and Faber, London.

*6

J. B. Birdsell

BIOLOGICAL DIMENSIONS OF SMALL, HUMAN FOUNDING POPULATIONS

With human populations striving to establish themselves in space as permanent communities, it is proper to look at some of the biological dimensions that can be identified as important in their founding phases. These factors include initial numbers, the biological intrinsic rate of increase, possible effects of inbreeding, and the optimal recruitment of the gene pool involved. It may not be too early to make preliminary sketches of the variables and their possible impact on this cosmic venture. Of necessity examples are chosen from appropriate population instances known on this planet. It is assumed that the major lessons derived from terrestrial examples can be projected into other environments.

The smallest initial number of space colonists is obviously two for bisexual humans, but this would represent a high risk and false economy. A much more reasonable figure would involve ten persons, presumably equal numbers of both sexes. But if maximizing fertility was important, fewer males and more females might be preferred. In any case, ten human colonists will provide an adequate number upon which to build future space populations.

The expansion of future populations from small founding numbers is best discussed in terms of the intrinsic rate of increase as defined by Birch. It is defined as the number of times the population multiplies per generation in an unlimited environment. Note this biological definition is very different from the like-named concept used by demographers today. The form of population increase in the early generations is a simple increasing curve, roughly comparable to the bottom half of Lotka's S-shaped curve. There are a number of human experiments in nature that bear directly upon the definition of the intrinsic rate of increase in our species. Three of them represent island populations sufficiently isolated so that the rate of their population growth is meaningful here. They have been reviewed elsewhere.

Perhaps the least satisfactory case involves Tristan da Cunha, a group of islands situated in the middle of the South Atlantic Ocean. They consist of five small islands of volcanic origin and are one of the most remote inhabited places on Earth. Of the islands, only the main one is inhabited. It consists of a single huge inactive volcano cone rising above 6,000 feet. An old lava plain about 2-1/2 square miles in area is supplied with running water and a good boat landing place. Here is the only permanent settlement on the island. The climate is temperate and markedly maritime. The flora is very limited and tough, consisting of evergreen bush, bog ferns, and tussock grasses. The vertebrate fauna is limited largely to oceanic birds, marine mammals, and fish. Cats, rats, and mice have been introduced by man. The original colony started in 1817 with a Scotsman, a Cape-colored woman, and their single child. Although the population history is not that of a closed community, the form of its growth curve was approximately logistic between 1817 and 1857 when the total number was ninety-five persons. The demographic history of the island as discussed by Munch reveals that during this phase the population growth rate approximated a doubling of numbers in each generation.

A better-known island community, as reported by Shapiro, is that of Pitcairn Island, originally populated in 1790 by six Englishmen and eight or nine Polynesian women. The climate of the island is mild, subtropical, and oceanic. The islanders in their 2 square miles of land area were supported by a primitive form of horticulture, primarily based upon Polynesian fruits and starchy tubers. Fish were occasionally available, but the meat from the few pigs was a rare item in the diet. Assuming a total of fourteen initial colonists, the population of this closed community during the first three generations increased at a rate slightly greater than a doubling in each generation.

A third island experiment involved the trihybrid population of the Bass Strait Islands between Australia and Tasmania. As studied by Tindale and Birdsell, this interesting mixed population originally consisted of eight reproductively effective white males and nine Tasmanian and four Australian aboriginal women. Economically, sealing was the original support base. In later years the exploitation of the seasonally abundant muttonbird, an oceanic petrel, provided meat, eggs, oil, and a yield of feathers, which were translated by export into a cash crop. The climate of the islands was temperate. The economy was supplemented by some marine products, an occasional kangaroo, and a meager form of horticulture. This group of twenty-one mixed people increased to 335 persons by 1945. Their intrinsic rate of increase, once the practice of infanticide imposed by the white sealers ceased, again was approximately twofold in each generation. At the beginning of the sixth generation, their genetic constitution approximated 22/64 Tasmanian, 6/64 Australian, and 36/64 European. So three insular cases of natural experiments under less than favorable circumstances demonstrated a doubling in population numbers in each generation.

The Australian mainland offers a historically documented instance of population increase in empty space in a most dramatic fashion. It was detailed by Cudmore in 1893. In essence, it relates that about thirty years before a young, aboriginal man had bolted from a white settlement at Popiltah Lake to the west of the lower Darling River. The account records that he had taken with him one or two young women and hid himself in a dense tract of mallee, which covers the country for about 600 square miles along the South Australian and New South Wales boundaries. There he carefully concealed himself from whites and blacks, living on kangaroo and whatever else he could get and obtaining water from the roots of the red mallee and the needle bush. For a period of thirty years he remained undiscovered and raised quite a sizable family. Except for the old man, who was very thin, all the others seemed very well developed and the young men were very swift runners. Norman B. Tindale has pursued the matter further with new evidence and concludes that the old man, Nanja by name, fled from the Danggali tribal country into the bush with two women. He has been able to reconstruct genealogical evidence indicating that each of these women bore ten viable children. This gives an intrinsic rate of increase of five times in a generation, a remarkable figure. The environment occupied by Nanja horde was empty of both aborigines and Europeans but was a most unfavorable one. They were totally dependent upon root water from two local scrub plants for survival. Their fertility, despite these disadvantages, was higher than that recorded for aborigines living normally under balanced ecological conditions. It rivals that of the Hutterites, the prosperous enclave of Canadian farmers in which the women at the completion of childbearing averaged more than ten children each. Obviously, both figures represent maximum value for our species. For the value attained by the Nanja horde to have been achieved in that desolate environment represents the worst scenario for human beings.

It is useful to calculate the rate of potential increase of human colonists in space under varying sizes of the founding population and their likely intrinsic rate of increase. If we assume that the intergenerational time is the time between midreproductive periods (approximately 25 years), then the population would have four generations a century. The assumption is based upon the onset of fertility at about 17 years of age and a prudent cessation of childbearing at 32 years of age. If an initial founding population of one man and one woman showed a doubling of numbers as its basic intrinsic rate of increase for eight generations, it would leave a descendant population of 512 persons. Given the more likely size of a founder population of 10 persons, at the end of 200 years they should leave 2,560 descendant individuals.

But if the data from the Nanja horde and the Hutterites are used, with a fivefold increase in numbers each generation, the numbers in the populations after two centuries become larger than needed for space colonization. With an initial seed stock of 2 persons of opposite sex, at the end of 200 years their descendants would number 156,250 people. Given an initial colony of 10 persons, they would have increased to a total of 3,906,250 individuals. The modeling of such figures is sufficient to indicate that human fertility is such that there would be no problem in generating sufficient numbers of people in space. The problem instead would be the construction of hardware and the provision of energy to support them.

Inbreeding has a variety of meanings, but we use it to refer to the mating between biologically close relatives. As a carryover from domestic herds and laboratory strains of animals manipulated to increase genetic homogeneity, the concept of inbreeding has been projected into the model-making of mathematical evolutionists in an unduly important sense. Many of the currently popular models investigating evolutionary processes have been formulated by Wright and an equally influential and parallel set of models by Malecot. Both of these workers have assumed that selection does not act on inbred lines of descent. This is a considerable error and introduces a profound bias into their conclusions and their measures of inbreeding intensity.

In fact, in real populations biological inbreeding is at most episodic, occurring at times of drastic population reduction and isolation. In general, very few individuals in a population are affected and the impact is negligible. In aboriginal Australia the issue has been compounded by social and cultural anthropologists who mistake the social rules of marriage as implying biological inbreeding where in fact they do not. The confusion arises from the fact that in a four-class aboriginal society marriages are arranged so that in terms of abstract kinship label, matings between first cousins are the preferred form. But those dealing with models of kinship have failed to realize that the Australian system involves classifying relationships and not biological ones. To illustrate, a preferred four-class marriage would be a man taking to wife his father's sister's daughter. This practice in a literal biological sense would be a first-cousin mating and would contribute to inbreeding if practiced long and perfectly. In actuality the system of classificatory kinship introduces many equivalents in the name structure of relationships. Thus, the man's father has many equivalent brothers in his own class or section, who in turn are labeled "fathers" of the man seeking the wife. In the same way, sister translates into many women of the same generation in the proper section identification. Finally, the so-called daughter of any of the these classificatory "mothers" is actually an equivalent "daughter" of the next generation rather than a biological daughter. In practice this system translates into an avoidance of mating with biological relatives. Tindale has pointed out that in arranging marriages great efforts are made to assure that the candidates have no known biological relationship to each other. This reality is so different from the model devised by kinship experts that there is small wonder confusion has arisen. In Australian tribes blood group data at the M-N locus indicate that the degree of heterozygosity shows no inbreeding depression in general and is what would be expected by random mating systems under the Hardy-Weinberg law. It may be concluded from this that in populations numbering 450 to 500 persons, of whom about 100 are effective breeders, marital choices can be so structured as to totally avoid inbreeding depression.

When it is further considered that inbred matings suffer from an unusually large amount of fetal wastage, the hazards of inbreeding even where practiced become reduced. In addition, in natural populations two parents may be replaced by two children as in the stable population models, but in fact four or five children are born and all but two eliminated. Thus, there is further wastage after birth. It is reasonable to presume that individuals carrying inbred genotypes would tend to be preferentially eliminated before reaching breeding ages. So the erosion of individuals of inbred origins would tend to eliminate inbreeding depression even if inbred matings were practiced in real populations. In well-established space colonies, matings could be so structured as to minimize the genetic consequences of inbreeding.

Granted that structuring of matings in a sizable space colony can avoid inbreeding depression, there still remains the problem of biological inbreeding during the early generations of the establishment of such a colony. Populations of surprisingly small size can manage this through careful avoidance of close relatives. A population of only several dozens of persons can avoid effective inbreeding at the level of first- or second-cousin matings if proper avoidances are consistently structured into the reproductive system. There are two examples from Australia that are pertinent to this problem. The first of these concerns three generations of father-daughter incest. It involves a camel drover who came to Australia from the Punjab, a province in northwest India, to manage the camels that were being imported for transportation in the Central Desert. Known locally as "Afghan," this particular man drove his camels out of the town of Oodnadatta in northern South Australia. He took to wife a full-blooded aboriginal woman who produced a number of his children. These first-generation offspring were F-1 hybrids and genetically contained 50 percent Afghan and 50 percent aboriginal contributions. In time this same Afghan took one of his daughters and began producing another family whose genetic composition was three-quarters Afghan and one-quarter aboriginal. He was a remarkable man and eventually took to bed one of the daughters of this generation, his own granddaughter, and she produced for him no less than eight children, six boys and two girls. This interesting family genetically was seven-eighths Afghan, derived from the original founding great-grandfather, and only one-eighth aboriginal from their great-grandmother. In theory this kind of extreme inbreeding should have resulted in marked physical degeneration evident in the great-grandchildren of the camel drover. But six of the eight great-grandchildren survived in adulthood and showed lower infant mortality than most families living under normal conditions in the center of Australia (fig. 6.1). This genealogy is by courtesy of Norman B. Tindale. The reason these highly biologically inbred individuals seemed to show no loss of vigor was, of course, that

Fig. 6.1. This is one of the great-grandsons of the man called Afghan. The great-grandson had seven-eighths of his genetic heritage from Afghan but showed no harmful effects of inbreeding, most likely because Afghan and his aboriginal first wife shared no harmful genes. (Photograph courtesy of J. B. Birdsell)

deleterious genes did not match up in the founding great-grand-parents. No doubt they were present in both Afghan and the aboriginal woman, but differed both in locus and in character.

A second example from aboriginal Australia is less well documented but just as interesting. In northwest Australia in the Great Sandy Desert lived a large tribe known as the Njangamarda. In precontact times they can be estimated as having numbered around 1,400 persons. They were divided into two subtribes, the northern Kundal and the southern Iparka. The investigations of Tindale uncovered the following interesting social dilemma. Both halves of the tribe were characterized by four-section systems. But the names and marital restrictions were such that the Kundal and Iparka could not exchange wives in accordance with the rigid system of their marital rules. They had not done so at least in the four generations before 1952. Tindale has hypothesized that this situation could arise only if a man married "wrong" and ran away into empty country to the north to escape tribal punishment. There he and his spouse founded a new population living by their own marital rules and in defiance of those of the Iparka to the south. A date cannot be placed upon this act of self-exile, but no dialect differences have had time to develop between the two sub-tribes. The Kundal at the time of contact numbered approximately 500 people. The point of interest here is that with a twofold rate of intrinsic increase, these numbers could have been achieved in empty space in eight generations. An examination of the frequencies of the M-N blood group phenotypes among fifty-two Kundal individuals showed an excess of 7.7 percent heterozygosity over that predicated by the Hardy-Weinberg equation. It would seem that from an initial founding couple, this population prospered and somehow avoided inbreeding depression. They demonstrated that it could be done.

A final consideration involves the recruitment of members for initial space populations. Concern here is not so much with individual physical fitness and vitality, but rather with genetic makeup. Obviously genetic screening would take place to ascertain through family medical histories that no harmful recessive genes occurred in the recent family lines of the candidates. Beyond that there are a number of serious maternal/offspring incapabilities that could be weeded out advantageously. For example, both in the ABO and in the Rh blood group certain types of matings tend to produce infant mortality. In the former this can be avoided by recruiting individuals with the O phenotype. An additional benefit would be that all members of the space colony would be both universal donors and universal recipients of blood transfusions in terms of this locus. The Rhesus locus also presents problems of incapability between mother and infant. These arise when the infant is Rh negative (d). This problem could be avoided by recruiting only individuals who were Rh positive (D).

There are in addition various polymorphisms that confer advantages in malarial environments but are disadvantageous to individuals in other environments. These include a variety of abnormal hemoglobins, a series of forms of the thalassemia, G6BD enzyme deficiencies, and others. No doubt by the time recruitment becomes active, the list will be greatly extended and the identification refined for individuals.

Even at the present level of genetic knowledge it would seem to be advantageous to recruit members of future colonies from a variety of widely spaced terrestrial populations. The benefits are several. Individuals from northwest Europe, peninsular India, and sub-Sahara Africa would show greater cumulative genetic differences in terms of normal polymorphisms than would crews recruited from more circumscribed populations, such as the fifty states of this country. There would be the advantage of maximizing genetic variability in future space colonies by this type of recruitment. In addition, just as Afghan and his aboriginal wife differed sufficiently genetically so that their harmful recessives did not match, so something of the same advantage would accrue in a founding population recruited to maximize genetic variability in this fashion. By the time of the actual recruitment, genetic engineering advances will no doubt allow vastly more types of genetic selection to optimize the gene pool of the chosen crew.

The biological dimensions considered here appear to offer no limitations affecting the successful recruitment of a founding human population in space and its continued growth there.

References

Birch, L. C. 1948. The Intrinsic Rate of Natural Increases of an Insect Population. *J. Anim. Ecology* 17:15-26.

Birdsell, J. B. 1957. Some Population Problems Involving Pleistocene Man. *Cold Spring Harbor Symposia on Quantitative Biology* 22: 47-69.

Cudmore, A. F. 1893. A Wild Tribe of Natives Near Popiltah, Wentworth, New South Wales. *Host Assoc. Adv. Science* 5:524-526.

Eaton, J. W., and A. J. Mayer. 1954. *Man's Capacity to Reproduce: The Demography of a Unique Population*. The Free Press, Glencoe, Ill.

Lotka, A. J. 1925. *Elements of Physical Biology*. Williams and Wilkins, Baltimore.

Malecot, G. 1969. *The Mathematics of Heredity*. W. H. Freeman, San Francisco.

Munch, P. A. 1945. Sociology of Tristan da Cunha: Results of the Norwegian Scientific Expedition to Tristan da Cunha 1937-1938, No. 13. Det Norske Vielenskaps-Academi I, Oslo.

Shapiro, H. L. 1936. *The Heritage of the Bounty*. Simon and Schuster, New York.

Tindale, N. B. 1953. Growth of People: Formation and Development of a Hybrid Aboriginal and White Stock on the Islands of Bass Strait, Tasmania, 1815-1949. *Rec. Queen Victoria Mus.* (Launcaston) N.S. 2:1-64.

Tindale, N. B. 1974. *Aboriginal Tribes of Australia*. University of California Press, Berkeley, Los Angeles, London.

Wright, S. 1921. Systems of Mating. II. The Effects of Inbreeding on the Genetic Composition of a Population. *Genetics* 6:124-178.

✸7

Kenneth W. Wachter

PREDICTING DEMOGRAPHIC CONTOURS OF AN INTERSTELLAR FUTURE

However venturous, an attempt to predict population into the long-run future can claim honorable antecedents. In his 1695 notebooks the grandfather of demographic studies himself, Gregory King, predicted the number of humans in the year 2000, arriving at a figure of 834 million, and continued the computation further, foreseeing 4.6 billion by the year 4000 and 9.3 billion in 8100, if the world "should last so long."

Predicting population into a putative age of interstellar migration, like any prediction, separates into two exercises—specifying future *transition rates* and specifying the initial state with which the prediction starts. Here transition rates mean rates of fertility, mortality, galactic migration, fission and fusion of communities, perhaps also of genetic transmogrification, and a description is required of the process generating these rates or at least of the bounds within which such a process is to operate. "Initial state" is not used here to mean the present, the population of Earth in 1985, but rather the short-run future, the initial configuration of extra-planetary settlement, with which any interstellar population begins.

The next few hundred years and the shape of any human communities first orbiting the Earth, adapting to the Moon, or sallying across the Solar System can scarcely be discussed without drawing on the kind of detailed knowledge of technological prospects that is outside a demographer's orbit. The next few thousand years, however, say down to the end of Gregory King's predictions in 8100 when travel beyond the Solar System might be contemplated, are in a different class. They invite speculations of such general tenor that considerations based on human demographic history may claim as much tentative relevance as any other approach. But thousands are not millions. Futures counted in millions of years demand the even broader perspectives of geneticists and paleontologists and a story shifting to new species, perhaps in great diversity, descending from or substituting for the human race. The years on which demographic lore might conceivably be brought to bear are middle years, years of interstellar journeying or of the earliest stages of settlement near other stars.

On the principle that general questions are more accessible than narrow ones, I will treat transition processes before initial conditions. But before either treatment, it is wise to ask whether they can be viewed separately at all. Are the dynamics of interstellar population largely independent of the early details of planetary departure? Or is all that follows determined by the circumstances prevailing as men and women, if they do, first leave the Earth behind?

In most respects, I believe, a good case can be made for the independence of future fertility, mortality, and migration from the initial state. A first argument is an argument by analogy. A characteristic property of most mathematical population models, both deterministic and stochastic, is the eventual independence, late in time, of important variables like growth rates or age distributions from the initial state. The so-called *weak ergodicity* theorems of Lopez, Cohen, and LeBras are strong results of this kind. Informally, such results obtain because there is substantial dispersion in almost every kind of demographic behavior, in the ages of childbearing both for each woman and among groups of women, in numbers of children, in ages at death, in propensities to migrate and, at another level, in next decade's rates for all these processes, whatever the rates are today. Unless extinction intervenes, populations come to ignore their pasts in periods of a few generations, essentially instantaneously when the time horizon stretches out to thousands of years.

The critical specific issue involving independence of initial state, which cannot be resolved by analogies alone, is the future course of fertility. Initial worldwide fertility decline would improve the chances of humanity to survive into a time when extraplanetary travel could become routine. Would such a pattern of low fertility, ingrained into new cultural values and economic and social arrangements, carry over into a long-term future, moderating the pace of population expansion across the Milky Way?

I argue that the answer is no: Fertility rates are too volatile and too little determined by those aspects of culture, economy, and society that perdure from generation to generation for us to expect that short-run fertility decline would imply continued low growth over epochs. This point is a controversial one. Certainly, no persistent fertility determinants are now known that enable prediction of total fertility rates even from decade to decade to within close enough bounds to tell the difference between net decline and substantial growth. The twentieth century record of attempts at such predictions in the United States is sad or funny, depending on one's mood. No major fertility swing has been consistently foreseen, with results that can be judged by the errors in total population predictions (owing mainly to fertility assumptions but somewhat to migration). Extrapolating from the projection record in industrialized countries, Michael Stoto considers the uncertainties in predicting the total U.S. population in the year 2000. He finds the uncertainties so great that his 95 percent confidence interval stretches all the way from 224 to 304 million. Perhaps, as many demographers believe, theoretical breakthroughs will soon prove fertility levels to be inherently predictable decade to decade after all. Perhaps the next round of predictions will avoid errors on the scale of the preceding ones. A less ambitious view, however, would be that the evidence weighs in favor of the possibility that those aspects of fertility, and its cultural, social, and economic concomitants that can be measured in the short term, account for little of the longer-term variance in fertility. If this hypothesis is tenable for decade-to-decade variability, then a far stronger hypothesis applies over millennia.

Those who reach contrary conclusions sometimes point to present-day fertility levels "below replacement" in the United States or Europe, and argue that modernized technological societies that make fertility the outcome of conscious individual choice and offer options for contraception and career opportunities for women are intrinsically low-fertility societies. On this view, the even more technological societies of space would be hard put to offer sufficient inducements to fertility to keep population sizes from decline. A careless version of this argument must be avoided. The observed fertility rates in the United States are "at or below replacement" only in the abstruse sense that if these rates persisted over several generations, then the size of the population, excluding migrants, would eventually stop growing. No low level of fertility rates in the United States has ever persisted over several generations. The 1981 U.S. rates that are called "below replacement" actually produced 1.7 million more births than deaths. Even excluding migration, the U.S. population was growing by 0.7 percent per year.

Far from declining to sustained low levels, fertility in the industrialized world since 1900 has fluctuated. The baby boom after World War II not only generated population growth at the time but also entailed population growth into the 1980s as the large numbers of baby-boom babies reached childbearing age. It is, of course, possible to claim that there will never be another baby boom. But far from being a truism, this is a prediction that the future of technological societies will differ from their immediate past.

Fluctuations in fertility of the size and frequency seen in the United States between 1930 and 1980 would produce population growth in the long term amounting to a tenfold natural increase every few hundred years. For instance, pairing the lowest crude U.S. birthrate between 1930 and 1980 in the Census Bureau's *Statistical Abstract of the United States*, 1.48, with the highest crude death rate between 1930 and 1980, 1.13, would give a tenfold increase every 659 years. Although these growth levels are far lower than those in the third world, they are more than sufficient to allow human expansion into space. These are not arguments that such expansion will occur, but only that it is by no means precluded by our current knowledge and experience of modernized societies.

Others who reject this point of view regard the baby boom as a short-term aberration in a long-term downward trend. The overall shift to lower marital fertility rates that divides the 1800s from the mid-1900s in the demographic history of most Western nations is not only assumed "irreversible," as it is often called, and replicable in the rest of the world, as is often hoped, but also unfinished. The forces producing fertility decline are seen to be still at work. This position would be easy to maintain if popular and intuitively appealing views of what those forces are could find better scholarly support. Unfortunately, causal explanations linking fertility decline to economic and social changes associated with industrialization or modernization, as measured by conventional aggregate indices, have fared poorly in the Princeton Fertility Project studies summarized, for instance, by John Knodel and Etienne Van de Walle. Different determinants seem to have been important in different cases, local cultural factors often eclipse general economic factors, and intangible effects of taste, orientation, and values come to dominate accounts. If the factors underlying European fertility decline were indeed of a type that cannot be predicted with any precision into the future, then they impose no strong constraints on space-age expectations.

It may be objected that demographic history provides numerous examples of patterns that persist for hundreds of years, sustained by continuing cultural systems. Three picturesque examples are Paul David's investigation of birth spacing in the south of France after the 1740s, citing the ideological heritage of twelfth-century Cathars, Peter Laslett's persisting localities of high bastardy rates, and Massimo Livi-Bacci's correlation of +.603 between an index of Italian marital fertility province by province in 1911 and proportions voting against divorce on the referendum of 1974. But these persisting patterns are differentials, not levels. The differentials are superimposed on changing overall levels. Birth spacing gets longer and shorter, bastardy rises and falls, Italian fertility declines, and the different groups change in tandem, their relative positions persisting but their absolute levels still displaying the general volatility of human demographic rates.

If this viewpoint is correct, it would suggest that any patterns of fertility level or timing worked out on the planet Earth as an accommodation to the impossibility of unlimited planetary population growth should not be expected to carry over to extraplanetary populations. The contrast between the North American colonial populations of the 1600s with their early marriage, high total fertility, and rapid natural increase, and the preindustrial European motherland populations with late marriage, moderate total fertility, and bounded growth could well repeat itself as a contrast between extraplanetary colonials and planetary homelands.

Assuming then considerable independence between initial state and long-term transition rates, what can be said about the bounds within which birthrates, death rates, and migration rates are likely to be generated over some interstellar future?

As for growth rates, let me urge what I take to be the long-term predictive content of the theories of the Reverend Malthus: Population size will track resource availability inescapably but at a distance. "Inescapably," inasmuch as a tendency for exponential population growth can outpace any expansion of resources that itself falls short of exponential rates. "At a distance," inasmuch as limitations on resources act on population growth, not only in the extreme by rationing subsistence but also well within extremes, by imposing a trade-off between growth and welfare. Malthus's "positive check," extraordinary mortality, can be complemented by his "preventive check," social arrangements that reward postponement of and forebearance in childbearing with individual prosperity and community sanction. The relative strengths of the positive and preventive checks dictate the distance between population size and available resources and hence, in the long term, the standard of living and quality of life that a society maintains.

It follows that all depends on two unknowns, the trajectory of resource availability and the standards of living that future societies will defend. It is easy to imagine episodes of sudden superabundance. Galactic voyagers happen on a habitable planet; a dozen generations follow of unbridled exponential growth. New alchemy learns to turn lead into oxygen; extraplanetary settlements double and double again. It is possible but harder to imagine progressive resource expansion driven by continual enhancements to technology achieving effectively exponential rates over thousands of years. An analogue of the subject of demography might be able to treat the growth of science and technology as a stochastic population process. But neither measurement methods, data bases, nor theoretical constructs is yet adequate to the task; whether the industrial and scientific revolutions are a one-time chance or an indefinitely sustainable process is still a matter of pronouncement more than of rational debate.

It is currently popular to interpret the billion year evolutionary record as a series of equilibria punctuated by short bursts of rapid speciation. It is also popular to liken the spread of ideas to the natural selection of genes; so-called memes become the genes of cultural evolution. Marrying these two notions, for whatever they are worth, would give a picture of sporadic increases and long plateaus of relative stability in the trajectory of resource expansion.

Further discussion of resources is best postponed to the concluding section on initial conditions for prediction. Whatever the resource trajectory, it is important to emphasize that some societies have managed to maintain a considerable distance between the actual population size and the maximum population size dictated by resource availability. Such societies have protected a standard of living well above subsistence, whereas others are recurrently exposed to

Malthus's positive check. The greater the premium on flourishing science or on social cohesion among spacefarers and the less the viability of coercion, the more might the advantage tip to societies that maintain a distance from resource limits. With high standards of living the definition of resources itself broadens to include the bases for individual privacy, variety, solitude, and uniqueness, resources perhaps more sharply bounded than those for subsistence itself.

Breathing room between population size and resource availability permits curves of population size over time quite different from the upward-sloping exponential curves and S-shaped logistic curves that pervade insect and animal ecology and theoretical population biology. A pattern of irregularly cyclic swings is just as characteristic of human populations as either the exponential or logistic. Such swings are apparent at the earliest period for which numerical estimates of national population, birthrates, and death rates become possible, after 1550 in England and slightly later on the European continent. The English preindustrial case provides data for an account by Ronald Lee of population, resources, and welfare that stimulates many of the speculations I am offering. So far as we know, the demographic dynamics Lee describes are of a kind that could have been sustained over thousands of years, a claim that could not be so safely hazarded for the faster than exponential growth of the last 150 years. I take the preindustrial experience interpreted in broadly Malthusian terms to be as plausible a model for populations in the next thousands of years as any based on the immediate human past.

Turning from growth rates to mortality and fertility, I put forward for the sake of argument a future regime of relatively high mortality and high fertility in contrast to present human conditions. I have in mind overall expectations of life returning below 50 years but with even greater variance in the distributions of age at death. This prediction posits an increase in accidental mortality associated with the sporadic failure of local life-support systems and with higher proportions of the population engaged in risky or adventurous tasks. It also posits greater vulnerability to epidemic disease from increased proximity. Medical eradication of diseases might proceed rapidly, but eradication of diseases would carry with it the eradication of immunity and might increase vulnerability to newly evolving strains. Finally, an increase in psychological stress should be foreseen, partly from the insecurity of interdependence, partly out of deprivation, as folk memory and personal experience separate farther and farther from each other among humans maturing out of contact with barnyard and forest, ocean, and an unconstructed wilderness of living things.

The higher levels of mortality in the future could easily be balanced by higher levels of fertility. History offers clear examples, like the Hutterites of South Dakota, where average numbers of children for mothers who survive to 50 can exceed eight or nine. In the case of the Hutterites, community support for the rearing and education of children lessens any need for trade-offs between quantity and quality of children on the part of individual families. This practice circumvents the pressure against high fertility that such trade-offs are believed to impose on high-status groups in developed societies according to the "new home economics." The same result might follow in space communities endowed initially with durable stocks of life-sustaining, high-technology equipment. Such societies might require little labor diverted from education into current production.

The biologically possible fertility levels instanced by the Hutterites would allow net population growth even with expectations of life as low as 20, especially if the age pattern of mortality, in line with the previous arguments, shifted away from infant ages and spread more evenly, if more severely, through adult years. Minor extensions of current medical capabilities could raise these maximal levels still farther, for instance, through drugs to induce twinning or to raise the proportion of females among births. The more speculative advent of genetic engineering, surrogate wombs for human fetuses in new animal strains, or test tube gestation might remove limits on fertility altogether. But it is not certain that they will make the old estimates of maximal fertility obsolete. The notorious vulnerability of complex systems, their frequent failures and costs of continual maintenance, may put tried and true modes of childbirth at a premium in space. In an artificial home the naturalness of childbirth may come more nearly to compensate its pain. Thus, the variety of achieved total fertility levels in demographic history, spreading across the allowable range and occasionally sustaining levels as high as the Hutterites over generations, is the best evidence that population growth will remain a practical human option if enabled by resource expansion and impelled by desire and by need.

The discussion then returns to the dominant unknown in future demographic dynamics, the trajectory of resource availability. The nature of resource development raises the one central issue on which independence between long-run demography and the initial conditions of extraplanetary expansion cannot be assumed. It makes a difference whether we are considering interstellar migration developing from a planetary population that in the short run continues to survive and serve as the base from which colonies spring or whether

extraplanetary migrants would early find themselves the lone
perpetuators of the race. We might phrase this question as the
difference between colony and ark. It is nearly certain that the
prospects of interstellar migration are not enough to offer any significant escape valve for the "population explosion," although the
cumulative long-term impact on an already largely controlled
planetary population of small continuing streams of out-migrants
should not be underestimated, as the recent calculations of E. A.
Wrigley and Roger Schofield for English emigration after 1620 suggest. But the short-run possibilities for creating small extraplanetary,
self-sustaining communities protected by distance from the risks of
extermination that now threaten the planetary population are real.
Such a strategy of diversification of risk in behalf of species survival
might become the implicit motive of an emigration enterprise.

Escape from oppression or catastrophe has been the driving force
behind many of history's most notable migrations, from the Diaspora
through the Pilgrims to the refugees from Nazi Germany and from
Eastern Europe after World War II. But historical examples of successful arks fully independent of the countries of origination and of new
neighbors are scarce to the point of nonexistence. Castaways on
Pitcairn Island and Tristan da Cunha are slender instances. The best
field for precedents would appear to be the Pacific migrations of the
Polynesians; in this connection the studies of Ben Finney assume
importance.

The superscienced space age civilizations that popular media
portray seem to require a colony picture, in which expansion only
follows on prolonged and successful development of new technologies. Interstellar migrants forced to make do with the very
minimum of technology that would allow long-term survival at all,
setting off barely in time to avoid being encompassed in planetary
catastrophes, call for a different imaginative portrayal. For them the
stocks of resources, oxygen, metals, and water might be severely
limited at the outset. Technologies to enhance these stocks, which
might be easy on Earth and require little serendipity in basic research,
might prove hard and slow in coming under the practical constraints
of an ark's endowment of equipment. From a demographic point of
view, such a regime of slowly expanding resources linked to medium
technology and entailing a slowly evolving population in space offers
the most intriguing possibilities.

There are happier scenarios for the origin of arks than the hasty,
crisis-driven adaptation of Solar System modules to stellar flight.
Dissenters unauthorized to draw on the extensive resources of official
civilization might save enough by private means to break away.

Explorers impatient with the pace or character of workaday technological extension might gamble on early maverick departures. The drive to pioneer might prove strongest in people with personal commitments to small-scale technologies. All these alternatives are eminently compatible with the possibilities for cometary nomads proposed by Eric Jones and Ben Finney.

The question of colony or ark impinges immediately on long-run population history through the issue of critical group size for group fissioning. In the colonial picture it is plausible to see interconnected centralized communities, of sizes as large or even much larger than present-day cities on Earth, with small-scale fissioning off of new community-founding groups supported at the periphery. In this case the problem of reducing risks of community extinction would have to have been largely solved independently of the processes of new community creation. In the ark picture, however, it is plausible to see much smaller groups subject to the continual threat of extinction. The minimum resource surplus necessary for mounting a new community would then be a larger fraction of the resources of the whole community. This requirement would slow the total pace of reemigration and would necessitate periods of population control. In behalf of long-term survival chances it would also lend high priority to the reduction of extinction probabilities by diversification through new community formation, and so it would impel the continuance of reemigration, even if the pace were slowed.

The demographic dynamics of arks give scope to far-ranging speculations. Suppose we assume a ceiling on the rate of expansion of community resources for an interstellar, independent ark community that is substantially lower than the 5 per cent year growth rates of which human biology is currently capable. Then it is possible to do calculations with branching process models that show that over substantial periods of time (the time constants depend on the community size assumed), the probability of either extinction or transgression of the upper feasible population limit becomes extremely close to 1, unless feedback between population growth and demographic rates occurs. This kind of calculation serves as a proof that demographic processes over long time periods are compelled to incorporate homeostatic adjustment; that is, excessive population growth or numbers lead to lower birthrates and/or higher death rates (and vice versa), producing a tendency for the population to stabilize itself. This conclusion holds for future groups of the size we posit for arks just as much as to prehistoric planetary groups. Thus, the form that homeostatic demographic mechanisms might take in interstellar communities assumes importance.

Interstellar communities would no doubt face more severe (if perhaps rarer) extremes of environmental conditions. Causes include occasional failure of community support systems, rendered more likely by their increasing complexity and their greater jeopardy to isolated or concerted destructive behavior under social and psychological stress. Hayden and others have emphasized that the effective limits to prehistoric population growth are to be measured not in terms of average availability of exploitable resources but in terms of the recurrence times of the most severe environmental conditions that groups can withstand. This claim ought to apply to interstellar as much as to paleolithic bands. Hence, the imperative arises for homeostatic mechanisms to avoid community extinction as well as for such mechanisms to limit growth in communities facing the risk of population outstripping the rate at which resource accumulation could support reemigration.

On the subject of extinction, it is possible to speculate as to how large minimum population sizes would have to be. A whole interbreeding population might well be made up, as in the hunter-gatherer groups described by Richard Lee, of a number of separate bands spread among different ships or comets, meeting occasionally and exchanging mates. If interbreeding populations were to be too small, "marriage squeezes" or their counterparts would strain the reproductive capacities of the groups. How small is too small can be calculated by computer simulation. Paleodemographic studies of this kind, following Wobst, have measured effects of incest taboos on population viability. Under hypothesized prehistoric demographic rates, the strains appear for interbreeding populations below a few hundred. Under rates like those hypothesized here, the strains would probably appear for still smaller groups. Low as these limits are, it seems to me that they might well establish the effective bounds on the rapidity of fission into new communities.

William Hodges, in the following essay, casts doubt on the viability of any such low populations, arguing that vastly larger groups are required if the core of human learning is to be transmitted from generation to generation. The problem is not information, which computers will no doubt store and retrieve in superfluity. The problem is understanding. The concept of a vanishing point passed out of human comprehension after ancient times, as Edgerton remarks, despite a passage in Vitruvius copied and preserved for some thirty generations up to the Renaissance rediscovery of linear perspective. It is possible to imagine the same fate befalling differential calculus, despite reams of texts stored in computer memories. Small interstellar bands, below 500 persons, would certainly need to cultivate values that would recruit larger proportions of the young into learned studies than any past societies that come readily to mind. Otherwise chance variations in the propensity for study over several generations could jeopardize the whole tradition.

Hodges also raises economic arguments against small band sizes, relating to the need to maintain "minimum efficient scales." I am inclined to believe, however, that wealth rather than income would be the important economic variable for interstellar migrants, so that current production would be less significant than the original endowments of the community. In that case the scale of production would not be critical. But this matter turns on the nature of future technologies. Endowed with a silicon chip maker, a band of interstellar migrants might be able to make a dozen chips at the same unit cost as a million. Such technologies would not demand thousands of consumers. By way of contrast, as Paul David of Stanford University has pointed out to me, technologies like present-day petrochemicals, which gain efficiency through volume, would require hundreds of thousands of consumers. On balance, nonetheless, I favor levels as low as several hundred for the lower boundary, which the trajectory of population size would have to avoid.

On the subject of the upper boundary, there is a particularly large literature on homeostatic growth-restraining population mechanisms focused on preindustrial Europe. Rightly or wrongly, this literature places great weight on inheritance. The chief mechanism for total fertility level adjustment of groups is seen to be the age at marriage. It operates in cultural contexts where marriage stands for initiation of childbearing and is restricted until resources are sufficient to be transferred to the younger generation, either from biological parents or through the opening up of new "niches" in the community. In small interstellar communities either some new homeostatic mechanisms would need to substitute for these or the requirements of homeostasis might stimulate customs of property, inheritance, household formation, and kinship-based social roles and intergenerational cohesion all more traditional than those which science fiction anticipates.

In summary, two broad possibilities for the demographic patterns of the long-range future are being contemplated here. First, at one extreme, for large colonies of high density to develop and gradually proliferate, mechanisms for combating the risk of extinction would take the form of reducing the variance of environmental conditions and truncating the extremes of the distribution of conditions. That postulates the solution of technological, sociological, and psychological problems before any principal period of galactic expansion and so presumes the long-run survival of the planetary population. Second, at the other extreme, small communities functioning as arks, escaping perhaps from the destruction of the Earth-based species, would be subject to continual risks of extinction. The prospects for survival would be based on the generation of sufficient numbers of small independent replicate communities, subject to high but relatively independent risks. The probability of local extinction of individual communities might be high, but the probability of simultaneous extinction of all independent replicates would be reduced. Long-term rates of total population expansion might be high, but in individual communities the homeostatic adjustments in fertility and growth that are thought to have characterized preindustrial European populations would reappear in some space-age guise.

In the short-term past, geographical expansion has usually followed a colony model; explorers, pilgrims, and settlers in a "new world" have maintained close, prolonged dependence on the surviving "old world" from which they came. In the long-term past, however, the Earth-based evolutionary process as we understand it appears to have incorporated a strategy of diversification of risk by the generation of large numbers of relatively independent trials. It tolerates high variance in conditions and high rates of local extinction, as an ark model would also entail. This reflection weighs analogically in support of the ark alternative as a prediction of the human future. Certainly, the specification of the initial conditions of long-term population processes in terms of a surviving remnant from a large catastrophe is deeply embedded in human myths. Perhaps these myths reflect genuine, if unarguable, understanding.

References

Cohen, Joel. 1977. Ergodicity of Age Structure in Populations with Markovian Vital Rates, II, General States. *Advances in Applied Probability* 9:18-37.

David, Paul, T. Mroz, and K. W. Wachter. 1983. Fertility Regulation in Natural Fertility Population. Paper read at Population Association of America, Pittsburgh, April 14, 1983.

Hayden, Brian. 1975. The Carrying Capacity Dilemma: An Alternate Approach. *American Antiquity* 40:11-17.

Hodges, William. 1983. A Response to "Demographic Contours." Paper read at the Stanford-Berkeley Colloquium in Demography, May 18, 1983.

Jones, Eric, and Ben R. Finney. 1983. Interstellar Nomads, Working Paper. Los Alamos Conference on Interstellar Migrations.

Knodel, John, and Etienne van de Walle. 1980. Europe's Fertility Transition, *Population Bulletin of the Population Reference Bureau* 34:6.

Laslett, Peter. 1980. *Bastardy and Its Comparative History.* Harvard University Press, Cambridge.

Lee, Richard. 1983. Hunters, Gatherers, and Colonists. Working Paper, Los Alamos Conference on Interstellar Migrations.

Lee, Ronald. 1978. Models of Preindustrial Dynamics with Applications to England. In C. Tilly, ed., Historical Studies of Changing Fertility. Princeton University Press, Princeton. Pp. 155-208.

Livi-Bacci, Massimo. 1977. *A History of Italian Fertility During the Last Century.* Princeton University Press, Princeton.

National Center for Health Statistics. 1982. Monthly Vital Statistics Report 31 (April 1).

Stoto, Michael. 1983. The Accuracy of Population Projections. *Journal of the American Statistical Association* 78:13-20.

Wobst, H. Martin. 1975. The Demography of Finite Populations and the Origins of the Incest Tabu. *American Antiquity* 40:75-81.

Wrigley, E. A., and Roger Schofield. 1981. *The Population History of England; A Reconstruction.* Harvard University Press, Cambridge.

*8

William A. Hodges

THE DIVISION OF LABOR AND INTERSTELLAR MIGRATION: A RESPONSE TO "DEMOGRAPHIC CONTOURS"

Tracking the Carrying Capacity

When a ship carrying a few pioneering settlers arrives at an uninhabited star system, the settlers will have to work hard to build habitations using the few tools they bring with them. After they have built places to live, they will have children, and in time the descendants of these original settlers will come to comprise a flourishing community. The purpose of this paper is to ask if it is possible to predict how the population of such a settlement will grow over time.

Travel by Starship

For a community on a voyage between stars, acquiring stocks of oxygen, metals, or water will be not *hard*, but *impossible*. When the travelers arrive at their destination star, their economic circumstances will change: water, metals, oxygen, carbon, and nitrogen will be available in effectively infinite quantities but not in an immediately usable form. The settlers will wish to establish space-borne green-houses, foundries, machine shops, habitations, and installations of many, many other kinds; the materials for making these things will be present in asteroids, but the settlers can afford to bring from Earth only the most basic and small-scale apparatus for retrieving and processing the asteroids. They might, for example, have only the equipment to retrieve very small asteroids, break them with hammers, smelt steel with parabolic, aluminum-foil mirrors, and forge small items using the tools and techniques of the medieval smith. (Indeed, in a number of ways the early communities may be more medieval in technology than in demography.)

This bootstrap process of building a space-borne society with the absolute minimum of initial tools will be very difficult. The settlers will have to start by making the most basic things, using tools brought from Earth. The things they make first can be used as tools to make other things, and slowly, step by step, the variety and quantity of output will increase. Clearly, the things absolutely necessary for life must come first; the requirements of comfort and safety, later. This means that for generations the settlers will be poor. They will spend their entire lives in conditions of extreme material deprivation and high risk of death. But these conditions do not mean a low growth rate. The settlement must achieve a minimum efficient scale as quickly as possible.

The Quality of Life and the Extent of the Market

Although a small community may hope to be self-sufficient in a few simple commodities, such as air, water, food, steel bar, steel sheet, and glass plate, any complex items can only be produced efficiently in large runs. One can easily list hundreds of items that a space-borne settlement will need: shoes, pens, paper, light bulbs, calculators, brushes, cups, computer tape or some alternative medium, space suits, clothes, blankets, caulk, salt—one could go on for pages and still not list everything. There must be at least a hundred items on such a list at least as vital as, for example, shoes. Thus, a community of only a few hundred, most of whom must work to produce the basic items, cannot afford to employ a specialist just to make shoes, especially since if you have a cobbler, you must also have an apprentice cobbler. On Earth microprocessors cost only as much as an hour of human labor, but to produce microprocessors efficiently, you need to make millions of them, so that the high set-up costs can be spread. A tiny population cannot hope to make microprocessors at all, let alone the hundreds of other things it will need. The concept here is a fundamental one in economics: Labor efficiency depends on the division of labor, and the division of labor depends on the extent of the market.

Lack of skill. The settlers can get along without a cobbler and without shoes, but for those specialties they must have, one person will have to cover what on Earth would be many separate specialties. A community of 3,000 might have one dentist and one minimal set of dentist's tools. A community of 300, spending the same proportion of its resources on dentistry, can only have a person who has nine other professions besides dentistry and one-tenth as many dentist's tools. Clearly, it makes sense for the smaller community to spend a greater proportion of its resources on dentistry—to have a dentist with only three other professions, even if one of the six professions given up is shoemaking, and to have a dentist's drill, even if this means not having a cobbler's last. Thus, besides giving up some specialties altogether, those specialties which a small economy does have will be handled with less skill. For example, one person might have to be doctor, dentist, pharmacist, and veterinarian. This person will not be able to spend as much time learning each profession as a person who has only one profession, and he or she will have to learn by apprenticeship, whereas on Earth a doctor or dentist can take classes from teachers who are more specialized in the profession. This doctor/dentist may be pretty good at routine things like delivering babies and filling teeth, but if an operation has to be performed that the doctor/dentist has never seen anyone do, then the patient will be in trouble.

Research and development. On Earth part of the set-up cost for many commodities represents the cost of research and development. The settlers will bring books and instruction manuals describing Earth's technology, so they will not have to do all the research themselves. But most commodities have set-up costs quite apart from research and development costs. Consider the example of shoes again: The demand for shoes in a tiny population will not be enough to keep a cobbler busy full time, so both the cobbler's tools and his expertise must lie idle most of the time. The cost of the tools and the cost of training the cobbler represent the set-up cost. A somewhat larger population can keep one cobbler busy full time, which is more efficient. A much larger population will have a shoe factory with specialized machines, such as a sewing machine for leather. Thus, the larger population will have to pay fewer hours of labor for each shoe, so there is economy of scale in shoe production which has nothing to do with research or development. Also, for some industries, a large-scale plant simply works better; a larger furnace or retort retains heat better because it has less surface area per unit of volume. Furthermore, the settlers will have *some* research and development costs, since they will have many unforeseen needs. They may find something unexpected in the star system, but a more important source of uncertainty is the development process itself: Things may work out quite different from what has been planned in the Solar System.

A machine to make microprocessors? Wachter suggests that future technology, such as a machine that simply puts out microprocessors using simple chemicals as inputs, would eliminate the economy-of-scale problem. But even if some profound technical developments could make it possible for people to survive in space indefinitely with a population size of only a few hundred, there does not seem to be any advantage in doing so. The living standard will always be higher with a larger population. Wachter's idea is that the migrants will be ill provided with equipment, and thus their population growth rate will be limited. However, he thinks the community will "fission," or split off, daughter settlements. But if they can fission, then they can also remain together and grow. To split, the community would have to reproduce each tool, and that might be beyond their capacity. It seems unlikely that an economy that can't make microprocessors will be able to make a microprocessor-making machine.

A lightweight machine to make microprocessors out of recycled garbage is a long way beyond our present technology, although technology will be much advanced before a starship leaves. However, our technology is not advancing in the direction of such a machine. It is true that our factories are becoming more laborsaving, but they are also becoming more specialized, more reliant upon specialized inputs, and *much* larger. To go from such a technology to a lightweight, self-sufficient, garbage-to-microprocessor machine would require a special development effort. To use such a machine would mean relying on technology that would not be tried and true, but ad hoc. Also, a ship hastily jury-rigged to escape a catastrophe or one sent by a small splinter group would not have such machines.

For most items it will require less weight, and thus be cheaper, to bring a stockpile of the item rather than a machine to make it. In the case of microprocessors, for example, it is hard to imagine a machine that could make a microprocessor weighing less than a million times as much as a microprocessor itself weighs, and a stockpile is much more reliable than a machine. But to bring even the stockpile from the Solar System will be horrendously expensive: By some estimates, with a flight time of sixty-five years about a hundred pounds of deuterium are required for each pound delivered to even the closest star.

The Population Trumpet

The population will be kept rigidly constant during the years on the ship, but after arrival the population will begin to grow at a high and exponential rate. A graph of population size would resemble a trumpet, with the years on the ship corresponding to the tube of the trumpet and the years after arrival corresponding to the bell. The advantage of rapid growth after arrival is that it minimizes the size of the stockpiles brought from the Solar System.

Consider the book you hold in your hand. A small initial settlement will not be specialized enough to have a paper industry, and to bring paper from Earth would be a prohibitive weight. Perhaps the settlers will have high-resolution computer terminals as thin as clipboards, so they will not need paper. But as long as there are only a few settlers, they will never be able to make such computer terminals or even paper. They can bring enough terminals for use on the ship, but during the postarrival years, as the population rises, the terminals will be scarce. They will have to sign up in advance to use a terminal for a few hours, and they will not be able to read or watch video (which could be done on the same terminals) every night. As the population grows they will someday be able to make not computer terminals but paper; they will print on the paper with the simplest technology, which is movable lead type. When the population becomes very large they will be able to make their own terminals, if they choose.

Thus, we can distinguish six phases in the economic development of a newly settled star system.

First, there is the time spent on the ship. Any tools or other items needed at the destination star will need to be brought from the Solar System; this means using energy to speed them up to the high speed of interstellar travel and then using more energy to slow them down at the other end. Furthermore, the fuel you used for slowing down must itself be carried all the way from the Solar System, which adds greatly to the energy cost. Because of this any item, a shoe, for example, which is used on the voyage but then thrown out might have an effective cost to the senders of the ship of ten pounds of deuterium, whereas the same shoe delivered all the way to the star might have an effective cost of one hundred pounds of deuterium. Thus, it seems likely that life on the ship will be relatively abundant and comfortable compared with the period immediately after arrival.

Second, there is a period of great scarcity after arrival before basic industry is established. The years after arrival will be lean years compared with the years on the ship. But in this second period the population will still be small, so the stock of super-high-technology durable goods (such as ultrathin computer terminals) will be plentiful.

The third period is after the basic industries such as air, water, food, glass, metal plate, and electricity have been established. Only the simplest things can be made in this period, and the output of these things will be low. For example, to make metal plate the settlers will need a crushing mill, a centrifugal furnace, several tanks, and a rolling mill, as well as mirrors to focus the sunlight. Suppose they have only brought one small furnace from the Solar System. The first batch will have to be crushed and forged by hand, and the first things they make will be a crushing mill and a larger furnace. Besides ore, they also need water, carbon, and electricity to make metal. These other inputs can be manufactured, but the tools to make these things all need metal. Thus, it will take a very long time before the settlers can build up to a large output, and most of what they produce must go to make more tools. The settlers can put off having babies during the second period, but by the third period they must have them, so there will be more mouths to feed per productive adult in this period. Nevertheless, the third period is a time of less privation and hardship than the second period.

The fourth period is when the population has grown enough so that the stock of sophisticated goods such as computer terminals brought from the Solar System is scarce. The settlers will arrive at their star in an extremely cramped ship. This will not be the ship that brought them all the way from the Sun, but only a piece of it since it costs too much to decelerate the whole ship. The stock of water per settler will be the minimum possible and the weight devoted to comfort as little as endurable. But everything on the ship and the ship itself will be a virtual miracle of lightweight design. By comparison, the habitations that the settlers build will be space shanties. The settlers will feel proud if they can rivet together something that will hold air. All the furniture in the habitations they build, such as cabinets, toothbrushes, stills to recycle urine, and ball bearings, will have to come from ship's stores. The stocks of these things will run low before the settlers are ready to produce them. This "stockpile depletion pinch" is felt most strongly during the fourth period. However, the basic industries will be producing more per capita and more per hour of labor in the fourth period, and some middle-complexity industries will be started, such as electric motors and metal fiber cloth. Thus, the fourth period will be worse than the third period in some ways and better in other ways.

The fifth period is when the settlement is self-sufficient but at a lower living standard than the Solar System.

The sixth period is when the settlement catches up with the Solar System in development.

How Bad Will It Be?

If the settlers cannot make an item themselves, then they must bring a great stockpile of it or do without. In the early years they will have to do without a great deal. We have seen that they cannot afford to have a full-time cobbler. Perhaps they will make some simple sandals, but I think they will go barefoot. (After all, there will be no outdoors to go to.) They will hope someday to have orbital greenhouses filled with cotton plants, or to make synthetic fibers from carbonaceous meteorites, but on the ship and for many years after, they will go naked or wear ultralight modesty garments from a stockpile. One day real food will be grown in the greenhouses, but in the early years they may have to make do with biosynthetics, fermented from their own feces by tailored microorganisms. They may not be able to afford full shielding against radiation. The artificial gravity may be weak and of the dizzying quality produced by the rotation of a small body; the air may be oxygen at low pressure; and they will have to spend their entire lives in habitations, which will be small, metal boxes. If they have a heavily female sex ratio to allow for high fertility, then the usual human sexual behavior pattern will be impossible. Privacy will be poor, as interior walls will either not exist or be tissue thin. Their political organizations are likely to become authoritarian, and the leader, lacking a brig worse than the prison they are all in, may resort to corporal punishment.

Perhaps the reality will not be quite so grim as this picture. Technology may alleviate some of these problems; for example, the synthetic food may turn out to be quite good, and the settlers will have all their lives to get used to it. It may be that even the lightestweight life-support system will involve the use of plants and domestic animals; if so, they may have sheep to provide milk, mutton, and wool. All the books, movies, pictures, and games of Earth will be available, as well as a powerful computer. Also, between now and the time the ships depart new commodities will be invented, and some of these will be sufficiently reliable and lightweight that they can be provided on an interstellar voyage. Perhaps the voyagers will say, "Our lives are hard, but at least we have it better than people did in the late twentieth century, before ecstasy music or the biofeedback brain game was invented." However, the invention of new comforts and luxuries between now and then may actually increase the relative deprivation of the voyagers. All the books and movies will show the settlers how little they have compared with the people back home. The first generation may regret, and the subsequent generations bitterly resent, that they ever left the Solar System.

To see the importance of extent-of-the-market effects on living standards, simply close your eyes and stretch out your hand until you touch something. Think about the first thing you happen to touch. Ask yourself: Could I make one of these out of an asteroid myself? Could I get along without it? Would the taxpayers of the Solar System spend a million dollars per pound to provide me with one? Whatever you touched, it probably took hundreds of people to produce, and millions of dollars worth of tools and factories. When one realizes how much specialization it takes to produce the simplest item, one can see how much privation will be felt by a population too small to specialize. We can imagine people living in tiny habitations made of hammered-out steel sheet caulked with mud, breathing badly filtered air, and eating artificial food. This picture may seem unrealistically pessimistic, but the conditions I have described are no worse than many people live under today—better in some ways, since I have assumed the settlers to be both healthy and adequately fed. If a fraction of the human species has for a relatively short time lived better than this, it is by developing a technology that relies on specialized machines and specialized skills.

The Cheapest Possible Spaceship

If an economy is wholly self-contained, making every single one of the tools it uses, and if it also maintains a high technical level, then it needs to have many people—probably millions. The ship is not self-contained; it relies primarily on tools brought from the Solar System, so it can have far fewer people. A ship's crew as small as ten is possible; there would be problems of lack of specialization, but it will be cheaper to overcome these problems rather than bring more people. A complement of ten might comprise

one medical person
one scholar-librarian
one person responsible for maintaining all the electronics

one person responsible for life-support systems
one person responsible for food production
one person in training to take over one of these professions
four children

These represent very broad categories. For example, the electronics specialist, who will also have to write computer programs, cannot possibly have the detailed knowledge to fix anything, given the enormous variety of electronic equipment on the ship. He or she will only replace entire modules when they break down, so the ship will have to carry a large store of spare modules, far more of each module than will actually be needed. Although ten might be enough to run the ship, it will not be enough to start the bootstrap process of building an economy after arrival—for that several tens or several hundreds will be needed.

The risk will be higher with ten passengers than with a hundred, but with a travel time of only sixty to a hundred years, the demographic risk—that is, the risk that the population size will go to zero through the random variation of fertility and mortality—will be small compared with all the other risks. If only women were sent, together with a sperm bank, the demographic risk could be made smaller still. The lower bound for ship population size will be set by the number of people necessary to bootstrap a technical, space-borne economy once the star is reached, rather than by any demographic factor.

The cost of travel is heavily dependent on the speed; by one estimate, to send 1 pound of payload to the closest star in 130 years requires 10 pounds of deuterium; doing the same thing in 65 years takes 100 pounds, and doing it in 33 years takes 10,000. Anything that the settlers will not need until a few years after arrival can be sent much more cheaply on another ship departing at the same time but traveling more slowly. It may be efficient to send the settlers themselves in several ships traveling at different speeds. The first batch of ten might do some basic task such as asteroid retrieval and processing. Ten years later, when there is a need for a larger number of specialists, another ship would arrive carrying both supplies and more settlers; in this way the Solar System saves fuel cost and the settlers save the trouble of feeding people before they are needed. Thus, the population of each ship might be as low as ten, even if the minimum necessary for bootstrapping is much higher. Note that although several ships are sent, this does not represent a long-term commitment to the settlement by the Solar System—all the ships leave at the same time.

A slow ark, which might take 1,000 years to reach the nearest star, would be much cheaper than a fast ark. But the number of suitable stars near the Sun is limited, so when a slow ark arrives at a star system, it will probably find it already settled. Also, if you had a certain amount of money and were planning to send a slow ark, you could invest your money at interest for a few years and send a faster, more expensive ark; it would arrive sooner. If you can earn 5 percent interest, then the optimal travel time is 80 years; if 3 percent, 100 years and if 1 percent, 175 years.

It may be that no one will volunteer to go in a slow ark, even among the billions of diverse people in the Solar System. An untouched star system represents a world of wealth; to go on a 100-year ark means that you and your descendants will suffer hardship for four generations but that the seventh generation will own mountains of iridium and gold. Also, if you go on a fast ark you will certainly earn a place in any history books written in the new settlement; they will call you a Founding Father and tell stories about you, like the stories about Miles Standish. So they may find volunteers for the fast arks; but slow arks are different. Going on a 1,000-year ark means subjecting forty generations to hardship, and the forty-first will almost certainly find the star system already occupied.

The Dependency Ratio

If interstellar settlement follows the "trumpet-shaped" pattern described earlier, with no growth on the starship and very fast growth upon arrival at the target star, then demography can provide insights into the problems that such settlers will have. In particular, the notion of the dependency ratio suggests a key difficulty. If the settlers hope to have a very high growth rate on arrival to reach minimum efficient scale as quickly as possible, this will clearly mean a high dependency ratio—that is, a large number of children to feed and educate per adult—just when the settlers can least afford this expense. Presumably, they will compromise by skimping on the care and education of the children; even so, the resource drain will be severe.

Between the beginning of deceleration and the establishment of habitations and basic industry, the settlers will work harder, suffer more privation, and take more risks than at any other time throughout the history of the settlement. These are the critical years. Although the end of the critical years is not marked by a definite event, we can take twenty years as an approximate figure for their duration. Surviving the critical years will be comparable to surviving a twenty-year stint on

Skylab; out of the question except for an athlete in peak physical condition.

The settlers will want to arrange the timing of births on the voyage so that during the critical years the dependency ratio will be as favorable as possible. The ideal would be to begin the critical years with only young adults; if that were possible, the population would consist entirely of productive individuals throughout the critical period. However, the productive years last so long that women who are adult at the beginning will be near menopause by the end. If they want to reproduce, they will need to have babies by at least the second half of the critical period. If at the start of the critical period, there are many young adults in the population, then the mothers of these young adults still will be alive. These mothers will be middle-aged but will become old by the end of the period. Time to care for these old women will be scarce. The best plan seems to be to start the critical years with females aged 5 to 25, with a few aged about 50, and a smaller number of males aged 18 to 35.

Plans for the age and sex structure of the population must take into account sexual needs. If you ignore sexual needs, it would be possible to have only females and a sperm bank, but there is no strong advantage to this, since you do not want to have the whole population pregnant at any one time anyway. An ideal would be to have an equal number of sexually mature males and females at all times. The plan I have given here has a surplus of men in the first half of the critical years and a surplus of women in the second half. Fifteen years after arrival, conditions will still be primitive, but at that time the settlers must begin having babies if the settlement is not to die out. At that time the youngest woman will be 20, and the youngest man about 33, but the first generation should be able to replace itself and more if it has a high fertility rate. Such large age-at-marriage differences are not unknown on Earth.

Thus, by carefully planning both the timing and sex of births, the dependency ratio can be made favorable during the most critical years. It is interesting to ask whether these elaborate population plans—planning what are traditionally private matters—will be obeyed and whether they will discourage people from volunteering themselves and their descendants for the trip. The plans are quite important. Refusing to make plans could double or treble the cost of the ship, since without planning half of the population at arrival could be dependents.

Since the arrangement outlined here has a surplus of sexually mature males over females in the early years, this might lead to a sexual dictatorship, with the captain dealing out desirable women to his lieutenants as one means of maintaining his power. It might be wise to put in the written constitution of the ship a provision that the captain must be a woman. Constitutions, of course, do not enforce themselves, but this type of provision might nonetheless be obeyed. The settlement will be divided into factions, probably two roughly equal factions, and each faction will have a leader whom it hopes to make captain when the present captain dies or comes to the end of her term. For one faction to chose a man in violation of the constitution would give an advantage to its opponents. It may also be a good idea to specify the age of the captain in the constitution.

I have tried to work out plans for the scheduling of births over time. These plans have to accomplish a number of objectives.

1. The population size at each point has to be appropriate to the degree of specialization required.

2. The dependency ratio has to be favorable at the most critical stages.

3. As each adult reaches the age of about 50, a child should be born to be the adult's apprentice and eventual replacement.

4. As far as possible, marriage partners should be provided for each person.

The plans are subject to two constraints. First, that for each scheduled birth there has to be a woman aged 15 to 45 to be the mother. Second, marriage partners should not be brother and sister or parent and child. The plans need to be flexible enough to allow for unplanned events such as early death, infertility, or someone taking an aversion to his or her only possible marriage partner. Accomplishing all these objectives turns out to be rather hard. For example, with a population of size ten there might be only two females under age 40. If one died or was infertile, then the other could produce the next generation, but these would then be all brothers and sisters. If so, they must either engage in incest or be celibate (using sperm banks for reproduction); either could produce social strains that would add to their other problems. My colleague Shelly Lapkoff suggests that an artificial uterus might be invented by that time, which would make birth scheduling easier. Another possibility would be to develop a technique to allow human fetuses to gestate in the uteri of sheep. The same technique would allow other species to be established at the new star without bringing along a breeding population of each.

Dependency and education. The arrival generation can teach their own skills to their children by the apprenticeship system, which is very efficient of adult time and also very effective—far more effective for most skills than learning out of books. But the second generation will be larger than the first and will want to acquire more specialized skills; thus, many people in the second generation will need to learn things from books.

For example, the ship itself may have only one medical person. Medical knowledge will be passed from master to apprentice three or four times, and perhaps one of these doctors will be a drunkard or a fool; it would be surprising if all four were good, able, sane persons. Let us imagine the sole medical man of the arrival generation. He is a combination doctor, dentist, and pharmacist. He is also his own pathologist and lab technician; he conducts classes in health and sanitation and washes the operating room table. With three professions to learn, he was not able to learn all of them fully, especially since his master was often too drunk to teach. Late in life he takes three apprentices. Since the population is growing, the settlement can now have three medical people instead of one, and it makes more sense for them to specialize. So he teaches his apprentices one profession each: a dentist, a surgeon, and a physician-pharmacist. Each of his three apprentices hopes to learn more about his or her own specialty than the old doc knows, and the only way to learn is from books. It is hard for them to learn just out of books when they are learning something the doc himself does not know, and there is no one they can ask a question. These young apprentices will also have to bridge the cultural distance between themselves and the profoundly different society that produced the books. Also, the doc's three apprentices were not the top students in their class by any means. The best students of the rising generation became astrogators or metallurgists or microbiologists—professions vital to the survival of the settlement; the doc's three apprentices were selected from about the middle of the class. Meanwhile, the doc is very busy; he has twice as many patients as he used to, half of them children. His three apprentices are children too, and not much help. The doc is facing the dependency ratio problem: more people per working adult. All the extra patients make the doc so busy that he does not get enough time to talk with his apprentices.

On ship there will be no shortage of adult time to spend with children, but the generation size will be small. Chance may easily produce an entire generation of dim students, who must then be the teachers of the next generation. Thus the cultural heritage of Earth will be transmitted through a few shipboard generations with small generation sizes and then through several poorly educated postarrival generations, so the level of knowledge may gradually decline. Earth's books, pictures, music, and video will always be available in compact form; but if the culture changes enough, later generations of settlers may not understand the ideas in the old, Earth books. If they form a new conceptual framework, they may be able to fit everything they read into that framework and not even realize that they do not understand what they read. The ship may maintain radio contact with Earth, but with at least nine years necessary to receive a reply to a question, this may not be enough.

Cultural difference from Earth is not in itself undesirable; but in the early years of the settlement, with memory of Earth's culture decaying and few or no scholars, artists, or scientists to be expected in such a tiny population, many of the advantages of culture may be lost. The settlers may misunderstand the technical and philosophical literature and the arts. They may lose the guidance which Earth's accumulated experience could have provided on such questions as how to repair space suits, how to govern a city, or how to arrange one's life.

The arrival generation will need to be very well educated, and there will be enough time to teach them on the ship, but they will need to know about building space habitations, locating asteroids, smelting metal—all things that will have no place in the lives of their teachers. All these things will have to be taught as abstractions by teachers who themselves have no direct experience.

The crew of the starship when it leaves the Sun may be composed of well-trained individuals selected for natural ability from the billions of humans in the Solar System, but it will be their great-grandchildren who arrive. The descendants of the original crew will have only the human average level of natural ability. The settlement of the stars will have to be accomplished by people of average natural ability and rather worse than average education.

Extrasolar Population in the Medium Run

After the settlers arrive at the star, their population should rise rapidly for several years. The population should rise to at least a million to allow for even minimal division of labor. But how much will it rise beyond that? I don't think there is any way to predict this, and indeed I expect that if several star systems are settled, their population histories will vary greatly after the initial period of rapid

rise. We can, however, list some of the factors that will affect
population.

For one thing, economy of scale will continue to be important even
when the population is very large. There are division-of-labor advantages to having a population of a billion rather than a million and even
to having a population of a trillion rather than a billion. Even when all
factories are operating at efficient scale, there is still the economy of
information. Once information—say a novel—is created, it can be
copied indefinitely without extra cost. With a larger population
people do not spend more time reading novels, but they can read as
many for less cost, since the cost of compensating the author is spread
farther. More important, they can have a greater variety of novels for
the same cost; a novel that appeals to only one in a million readers can
still have enough sales to be profitable.

A population of less than a billion should not face a resource
constraint; there should be unused asteroids very nearly as good as the
ones used by the original settlers. But although there may be room to
grow in purely economic terms, socially there may be costs to growth.
If, as I expect, almost everyone will be living in a single metropolis,
then additional people will mean greater congestion and a change in
the quality of life for everyone.

In predicting the course of population history, we need to consider
not only the number of children which each couple will desire to
have but also what population size will be considered desirable by the
community as a whole. It is the community as a whole, rather than as
individuals, which benefits from economy of scale and suffers from
congestion. A general social consensus on the desired population size
can affect individual fertility either by taxes or subsidies on large
families or by a social norm favoring large or small families.

To decide how the society as a whole will weigh the costs and
benefits of a large population, we want to look at the relation between
a given settled star system and the other star systems, especially the
Solar System, in the human sphere of settlement. Although a delay of
ten or twenty years to receive a reply to a question would seem
intolerable to us, accustomed as we are to instantaneous communication, we can expect that human institutions will adjust to this situation. We might expect that an inventor or an author would count on
the royalties paid by other star systems, even though it would take
more than a lifetime for a message to reach those star systems and
return; a long-lived institution such as a bank would be willing to pay
him the discounted value of the royalties to be received in the future.
The distant star system would have to pay royalties in the form of

information. Since each star system would in some ways be a part of the much larger human sphere, the cost of having a small population, in terms of the economy of scale in information, is not as high as it would be if the star system were isolated. However, there may well be competition between the systems. The wealth and power of each system would consist of information: inventions, works of literature, other works of art—whatever could be sent by radio. Since the supply of such information would presumably be proportional to population, there would be an advantage to each system in being large.

The Long Run

The long run may be more predictable than the medium run. It is not certain that even one ship will be sent from the Solar System to the stars. If one ship is sent, and a new civilization is established in another star system, then that civilization may never send out a ship of its own, and the Solar System, having sent one ship, may never send another. Lots of things could happen to prevent the expansion of the human species out into the Galaxy—there could be a nuclear war before you finish reading this essay, for example. But suppose the expansion does get started. Suppose ten or twenty star systems are settled by ships sent from the Solar System. At that point the expansion process could not be stopped. Cultures differ—if there are ten or twenty star settlements, then at least one will be chauvinistic and feel that its culture is so superior that it should be spread to other stars. If there is even one star settlement that believes in reproducing itself, then it will reproduce itself, and soon there will be many star settlements that believe in reproducing. There will be a kind of selection process: The star system with the qualities necessary for reproduction, whatever these qualities might be, will reproduce, so that at any later time most star settlements will have whatever qualities lead to reproduction.

One quality that would lead to the reproduction of a star culture is a conscious belief, rationally applied, that reproducing is a good thing. Thus, we may expect that star settlements will compete to establish as many daughter settlements as possible. It does not matter if there is no logical reason for wanting to have a lot of daughter settlements. People can believe illogical things, and those who logically refrain from sending out ships will simply be left behind as the reproducing star civilizations become more and more numerous. We can expect that the expansion to the Galaxy will occur even if there is no good reason, such as overpopulation at home, for it to occur.

Suppose you live in a star system deep within the interior of the sphere of human settlement and you want your star culture to have as many descendant civilizations as possible. What is the rational thing for you to do? I believe the best thing to do is to spend your entire budget on just one ship. Spend enough on fuel to make that ship very fast and build a life-support system that will last thousands of years. Then send the ship all the way to the galactic core. At the core, stars are closer together. By planting a single settlement at the core and letting it grow exponentially, your star system will have far more descendants than if you sent many ships to stars just beyond the edge of the sphere of human settlement. If this is the best strategy for such a system, then humans will spread from one edge of the Galaxy to the other at the full speed of the very fastest ships they can build.

Although the competition for reproduction should lead to humans spreading across the Galaxy, it is possible that they will never fill in the Galaxy, occupying every star within it. It may be that there are certain classes of star systems that are uneconomic to occupy; perhaps stars in star clusters are more desirable because in such a cluster stars can communicate with one another with shorter waiting times. It might seem that given billions of inhabited systems, there would have to be at least a few that would choose to send starships even to low-quality stars, particularly after millions of years. But the Galaxy may in some ways be one culture, in which case a star that is considered undesirable by that culture will never be inhabited—just as snails in England are never eaten. It seems hard to imagine the Galaxy being one culture, given the length of time it takes for a message to cross it; but as the human race grows older it may develop longer-lived institutions, allowing investments to be made even though the return on those investments will take thousands of years.

Since communication is limited by the speed of light, it may often happen that a ship is sent to a distant star only to arrive after hundreds of years to find the star has been settled in the meanwhile. Since star systems compete to make daughter settlements, they will have to send fast ships at prodigious expense for the sake of getting first to a star. To prevent this wasteful rivalry, humans may try to agree on the allocation of stars before the zone of human expansion becomes too large. Perhaps when the first starship sets out it will be agreed that a particular 2 percent of the Galaxy will be reserved for the new settlement and its daughter settlements, while the other 98 percent is reserved for the Sun and its other daughter settlements. Naturally, there would be a certain amount of star theft. There could even be property rights in distant, unsettled suns; they might be bought and sold—if so, they would be quite cheap.

Although star settlements will try to have daughter settlements, there is no particular reason to think each settlement will try to have as many inhabitants as possible. One settlement may decide to intercept all the energy of its sun and use all the energy to send out ships. But although there is a selective pressure toward reproduction, the whole Galaxy may be filled in before anyone becomes that fanatical about reproducing.

The different star settlements will compete with one another to have as many daughter settlements as possible, and in this competition there will be a large payoff to the first settlement to establish a settlement in a Galaxy other than this one. Well before the Milky Way is in any sense crowded, a ship might be sent across intergalactic space.

Reference

Heppenheimer, T. A. 1977. *Colonies in Space.* Stockpole, New York.

9

Eric M. Jones and Ben R. Finney

COMMENT ON HODGES'S "THE DIVISION OF LABOR AND INTERSTELLAR MIGRATION"

Distilled to its essence, Hodges's question is this: How small could a stellar settlement be, especially during the early years? The problem is bootstrapping. The initial party of settlers would come well supplied with stores of ready-made goods, but how would they build the economic base to support future (and much larger) generations? How would they pass needed skills to their children?

In simple terms we believe the answer must lie in computers and robots. If interstellar migration is to happen at all, human skills must be supplemented with capabilities stored in machines. We are in complete agreement with Hodges that a few hundred people, equipped with 1985 skills and technology or even twenty-fifth-century skills stored in human skulls would be in for a very difficult time. Interstellar settlement based on an intricate technology would not be easily maintained or propagated by hand. Capable (intelligent?) machines will be essential.

A proper response to Hodges requires an assessment of future developments in information processing. Will human-machine complexes be up to the task? What would be the level of human participation at each stage of the growth of the technology base that must be put in place to supply the community's needs as the ship's stores are depleted? Could machines be effectively used as teachers of new generations of humans? We can only suggest some trends.

Tools and skills will, of course, be vital to the survival and growth of a stellar settlement. Living in space is a high-tech enterprise. At its simplest, technology provides single-level tools and skills. For example, a chimpanzee shapes a stem of grass and uses it to extract termites from a mound. The hand selects and shapes the grass stem (skill in toolmaking), inserts it into the mound (skill in tool use), the termites are eaten—end of process. Very early in hominid evolution our ancestors carried technology beyond such one-level processes. Although one might argue about whether the fashioning of stone tools by banging two rocks together is one-level technology or two, shaping spears, making cooking fires, and fashioning clothes from animal skins requires multilevel technology. Tools are used to make other tools. In the modern world most technologies are many layers deep, and each level requires its own skills.

Hodges's argument is based in large measure on this depth of technology, especially in the high-tech world of interstellar migration. In his example of making shoes, the prospective cobbler would need not only a last but also the equipment to mine appropriate ores from asteroids and to make leatherlike material as well as nails, thread, or glue. Shoes (not to mention microprocessors) are very complex artifacts when the cobbler must start with an unprocessed asteroid. If each stage of the process had to be done by hand and human sweat, the interstellar settlers would not go barefoot; they would die.

Fortunately, our multilevel technology has produced more than skill to make tools with which to make more sophisticated tools (a lathe in the hands of a skilled machinist is probably the height of such a layered technology). Very recently, technology made a transition to a new level of sophistication. Before the Computer Revolution each layer of technology had to be added by hand. Human intervention was necessary at each step. It is this type of bootstrapping to which Hodges refers. Computers add a new dimension. They are not merely tools with which to make more sophisticated tools but rather skilled tools that can supplement or even replace human intervention in the development of the layers of technology.

We suspect that the primary task of interstellar migrants will not be making shoes or microprocessors or pounding bits of asteroidal material to be fed to smelters. Rather, we suspect they will be troubleshooters—caretakers of computers and robots that will do all the basic tasks. We doubt there will be any flesh-and-blood miners or cobblers. Instead, the human settlers will oversee the operation of a complex of machines that will mine asteroids, produce stores of processed materials (metals, foamed glass, and all the rest), and most important, make machines to make the final artifacts including shoes and microprocessors.

Perhaps we are asking too much of machines. We think not. The possibilities of storing skills in computers have been exploited only on the most primitive levels so far. Contemporary computers are balky, uncooperative (they really are idiots, repeating erroneous programming endlessly), and much in need of human intervention. But we should remember that the Computer Age is only a few decades old. Already there are hints of the next stages in which computers will become more cooperative partners in human endeavors. The reader is urged to consider plans announced by Japan to develop in the next decade truly skilled machines (see *The Fifth Generation* by Edward Feigenbaum and Pamela MacCorduck). If only a fraction of Japan's announced goals are achieved (by Japanese, Americans, or whoever), we should soon see machines capable of many of the skills needed for interstellar migration.

One development already under way is that of the so-called expert system—a type of computer program that has built into it both an extensive data base and a sophisticated decision logic that enables it to operate at the level of a human expert. Such expert systems are developing into valuable tools in chemical research, medical diagnosis, and oil prospecting, and we think that they and their descendants will be invaluable to small, spacefaring populations.

Let us imagine how an expert system might answer one of the problems posed by Hodges. In a small space settlement a single person, Hodges argues, might have to be doctor, dentist, pharmacist, and veterinarian. This lone individual, Hodges continues, "will not be able to spend as much time learning each profession as a person who has only one profession, and he or she will have to learn by apprenticeship, whereas on Earth a doctor or dentist can take classes from teachers who are more specialized in the profession. This doctor-dentist may be pretty good at routine things like delivering babies and filling teeth, but if an operation has to be performed which the doctor/dentist has never seen anyone do, then the patient will be in trouble."

Reality might go something like this. The ultimate general practitioner will be skilled in the routine procedures, but more important, will have a good working relationship with a computer-based partner. Between them they should be able to come up with an excellent diagnosis. After all, even the best of terrestrial specialists must go to books and colleagues for help with the nonroutine cases. The computer half of the team would have access to a superb medical library and be able in minutes (rather than hours or days) to present relevant information to the GP.

We might expect that before a tricky operation the GP-cum-surgeon could review detailed (probably 3-D) records of similar operations performed by Solar System experts. The computer would closely monitor the operation and the patient's condition, offer advice or even replays of the crucial stages, and help analyze unexpected conditions discovered during surgery. The key ingredient, if the interstellar GP is to keep the community healthy, is access to information. If current trends in information processing are any indication, the interstellar settlers will be well served.

Modern civilization has produced a vast store of knowledge. Finding and digesting the relevant parts can be quite a problem. Knowledge is useless if it can not be accessed. A typical modern-day specialist has an office full of unread journals. There have probably been cases of competent doctors losing a patient because a description of a likely cure lies unread on a desk. The coming phases of the computer revolution offer a means of managing this mass of data. We will need to become adept at using information systems that can help us extract the useful bits from the accumulating mass of data; we will need to learn some new skills.*

---

*A medical colleague, Dr. Jon Johnson, has suggested to us that the main concerns (in increasing order of severity) of the ultimate general practitioner will be childbirth, accidents (fractures, burns, etc.), degenerative diseases, and infectious diseases. As in any pioneering venture, the risk will be greater than in modern society. Acceptance of that risk will mean that the group must plan for the possibility of losing members to accidents and disease despite the ultimate GP's best efforts. With sufficient backup from the information processing systems, the loss of a single individual should not be catastrophic. In fact, should a group be so vulnerable, they ought to think twice about leaving the Solar System in the first place. Of far greater concern would be the diseases that could threaten the entire group: degenerative diseases like arteriosclerosis and diabetes, which could contaminate the gene pool, and the whole range of infectious diseases that will invariably go wherever we do (see Alfred Crosby's essay). The most important role for the ultimate GP and his computer partner (after childbirth) may be that of medical detective.

Human beings haven't changed much since *Homo sapiens* first appeared tens of thousands of years ago. Collectively, we know more than our ancestors did. But individually, we face the same problems. Which bits of collected human wisdom must we master to make life possible and fulfilling? It can be argued that a woman gathering roots in the Kalahari Desert is as much an expert as a nuclear physicist or the anthropologist who records her life story. The skills she needs to survive, to produce food from one of Earth's harshest environments, are intricate. Few of her "civilized" contemporaries could survive in the Kalahari. They simply would not have the right skills.

We suspect that the skills vital to the survival of an interstellar settlement will be equally intricate. A Kalahari woman knows how to read the telltale signs that indicate a succulent root lies buried a few inches below the surface. Her interstellar cousins must know how to "read" a computer-based knowledge system, how to troubleshoot, how to learn. These are ancient skills in new clothing. Hodges raises some important issues. An interstellar settlement would be faced with the need to bootstrap on a massive scale. We propose that a successful solution based on information processing will be found.

SECTION III

Migrating Societies

INTRODUCTION

As beings conscious of our past and anxious about the future, we often look back when we try to think ahead. Those whose passion is imagining how we will colonize space frequently are drawn to earlier episodes of human expansion—for models, insight, or just inspiration. Yet when examples from our migratory past are cited, typically they are only mentioned in an anecdotal, unsystematic way. For example, although the Polynesian settlement of the Pacific is often invoked as a dramatic precedent for expansion into space, the full implications for space colonization of the saga of small groups of settlers sailing into unknown seas to found colonies on tiny volcanic islands are not explored. In another vein, advocates of space migration often extol the courage and foresight of Columbus and Isabella and declare that the opening of new worlds in space will bring benefits to humanity akin to those that followed the European discovery of the Americas—without reference to the fact that this terrestrial New World was already inhabited or to the devastation that Europeans and their diseases visited upon the indigenous Americans. If the anthropological and historical record is to help us think about the human side of space expansion, it should be examined more critically and systematically than heretofore has been the case.

"Those who cannot remember our migratory past will be condemned to repeat its mistakes as we expand into space." This adaptation of Santayana's famous dictum might seem overly dramatic. But given the harsh environments that will be encountered in space, colonizers can hardly afford to ignore any insights or lessons that might be realized through examining how migration took place here on Earth, the structure of the small groups that actually migrated, and the evolution of the colonial societies which resulted.

Of all the past migratory episodes known today, the Polynesian expansion into the Pacific is the one that, as Ben Finney points out in the next essay, so often attracts the attention of those thinking about space migration. Despite the immense gulf that separates Stone Age Polynesians from those who will migrate into space, a kinship is there. To make the equation, substitute double-canoes for spaceships, the immensity of the Pacific for the vastness of space, and uninhabited volcanic islands and wave-washed atolls for empty planets, asteroids, and comets. Especially intriguing is the structure of the original migratory society and the examples of societal change that followed as the many separate islands were settled and the colonial societies evolved in comparative, if not absolute, isolation. On the larger islands, such as those in the Hawaiian chain, we can see societies growing rich and complex, whereas those on the smaller islands remain small and spare, confined by the limited area of a narrow atoll ring or a minuscule volcanic peak thrust out of the ocean. The Pacific provides a natural laboratory for the social dynamics of our migratory species.

In fact, we would argue that studies of Polynesian societies and of other small, isolated populations around the world should be required reading for those who one day will plan the first human settlements in space. If current dreams of founding colonies on the Moon, on Mars, or in artificial biospheres constructed in space itself are ever to become a reality, careful thought will have to be given about how to structure colonizing populations and organize human relations within them. There is a precedent. A generation ago when it was realized that plans for space flights of long duration raised problems of efficiently structuring human interaction in the cramped, hazardous, and isolated microenvironments of space capsules, a rash of studies were commissioned to examine the sociopsychological problems encountered by the crews of nuclear submarines and of scientists wintering over in Antarctica. More recently, this research has been extended to the crews of oil tankers and oceanographic ships. Although such studies may be useful for the current stage of preliminary space exploration, they are too limited to be of much value when it comes to thinking about founding permanent colonies in space. These analyses of small, overwhelmingly male groups working together on brief and narrowly defined missions usually under military or naval discipline need to be extended by studying the structure and organization of isolated and self-reliant breeding populations, of, in other words, real communities composed of men, women, and children.

Of course, the most basic human community may be the first one to have developed: the hunting and gathering band. As Richard Lee emphasizes in chapter 11, until the invention of agriculture all humans lived in such small and mobile communities. Furthermore, although the particular group Lee discusses, the !Kung San of the Kalahari Desert, is not now migratory, he makes the additional point that such small groups were the ones in which humanity originally spread over the Old World and then across the Bering Straits into the Americas. The hunting and gathering band is therefore a proven type of social organization with a long track record of migration—which, along with its minimal size, is the reason why we chose it as a model for the smallest human community that could colonize the comets.

With the other two groups Lee discusses, the Greeks and the Vikings, we enter the realm of history proper. Abundant documentation, as well as archaeological remains and migratory sagas, inform us about their motivation for colonization, how they organized their moves, and the fate of the overseas settlements that resulted. In chapter 12, Ben Finney continues this examination of maritime expansion in the historic period by focusing on the European Age of Exploration and the immediately preceding, but stillborn, Chinese initiative to sail to far-off islands and continents. Why the poor, technologically backward Iberians were to be the ones to open up the world ocean for trade and colonization, and not the rich and technologically advanced Chinese, is one of the puzzles of world history. As Finney's essay suggests, trying to answer that question and others suggested by this extraordinary contrast in basic attitudes toward exploration and colonization can lead to some interesting insights into the possible political and economic dimensions of the first stages of space settlement.

When human communities are well established in space, how will they relate to one another? One can conjure up all kinds of political scenarios yet overlook a basic biological one. As Alfred Crosby asserts in chapter 13, after the hostile natural environment, the second most dangerous enemy in space will be ourselves and our diseases. Crosby illustrates this by recounting the horrendous consequences to the indigenous Americans of the intrusion of Europeans and their microbes and viruses. If major fractions of isolated Earth populations could be wiped out by diseases nurtured elsewhere on the globe, what will happen to communities long isolated in some far sector of space who are suddenly visited by fellow, but nonetheless alien, humans bearing a separately evolved microlife? Actually, this situation is already upon us at a sublethal level. After wintering over in Antarctica, scientists often succumb to respiratory infections with the arrival of the first visitor of the spring, for it does not take lengthy isolation to be rendered vulnerable to the latest variety of common cold, flu, or worse. Let us hope that progress in medical science keeps pace with our ability to spread throughout space.

The colonial takeover by European powers of already populated lands might seem irrelevant as a model for space expansion, unless one imagines neighboring worlds populated by intelligent but technologically infantile species. Yet if we do disperse widely through space, marked cultural and technological differentiation is sure to follow quickly, setting up a situation in space not unlike that which led to Earth's recent era of European imperialism. In chapter 14, Nancy Tanner offers a nonaggressive model for intercultural space travelers: the Miningkabau who stem originally from the island of Sumatra. This Indonesian minority appears to be a highly adaptable people who traditionally traveled widely in Southeast Asia (and now the world) in search of adventure and fortune, but in anything but an imperial way. They seemingly are content to fit in wherever they can, adapting to the host culture while retaining their own identity.

As Tanner's discussion of the Miningkabau and their continued penchant for travel suggests, the search for terrestrial precedents should include studies of contemporary migration as well as long-past examples. In chapter 15, Douglas Schwartz briefly examines a broad range of contemporary communities that have resulted from recent migratory movements: for example, American farmers who moved from Texas to New Mexico because of drought and economic depression; East Indian peasants recruited to work on Fijian sugar plantations; and Doukhobor farmers who migrated from Russia to Canada for religious reasons. This preliminary study suggests some variables that appear to be critical to migratory success and points the way to a comparative sociology of migration that might one day be a required subject for all space settlement planners.

The final essay of this section is not by a historian or an anthropologist but by a philosopher, Edward Regis, Jr. In it he raises, and to his satisfaction solves, a moral issue: Do space voyagers on multigenerational interstellar arks have the right to reproduce and hence condemn their offspring to exile from the home planet—first in the ark and then on some alien world? Although space enthusiasts might grow impatient reading such a painstaking analysis, it does alert us to the fact that manifold moral issues and moral decisions await us in space.

"Civilization is a stream with banks. The stream is sometimes filled with blood from people killing, stealing, shouting and doing things historians usually record, while on the banks, unnoticed, people build houses, make love, raise children, sing songs, write poetry and even whittle statues. The story of civilization is the story of what is happening on the banks. Historians are pessimists because they ignore the banks of the river."

—*Will and Ariel Durant*

"Only a lifetime ago, parents waved farewell to their emigrating children in the virtual certainty that they would never meet again."

—*Arthur C. Clarke*

* 10

Ben R. Finney

VOYAGERS INTO OCEAN SPACE

Perhaps the most evocative phase of humanity's spread over Earth for thinking about the future expansion into space is the exploration of the sea and the foundation of island and overseas colonies. Like space today, the ocean was once a vast unknown. Just as we cannot definitely say if any planets suitable for life orbit other stars, so our not-too-distant ancestors did not really know what lands lay over the seas. Then, when those first intrepid voyagers took to the sea, like the space pioneers of today they had to develop a revolutionary technology to penetrate an alien environment. And those colonists who followed in their wake had to uproot themselves from all that was familiar and take their chances on foreign shores.

The one episode from our species' seafaring past that haunts the minds of those who dream of colonizing empty, virgin worlds is the prehistoric expansion of the Polynesians into the Pacific. Despite its Stone Age setting, this Polynesian experience would seem to have all the basic elements of a rousing good tale of human migration applicable to space as well as to the sea: an adventuresome people who developed a new technology to penetrate far into the unknown where they discovered and settled islands never before touched by man. With the coming expansion into space in mind, let me sketch what we know of this extraordinary migration and of the many generations of seafarers and colonists who actually braved the sea and founded new settlements on previously uninhabited lands.

The Polynesian story begins some 10,000 years ago along the shores of the southeastern end of Asia. There, as sea levels were rising at the end of the last glaciation, some enterprising people learned how to adapt to the changing coastal environment by combining newly discovered techniques of plant cultivation and animal husbandry with innovative ways of fishing and sailing in offshore waters. These coastal farmer-fishermen used their canoes to spread along the shores of what is now Vietnam and south China, to offshore islands including Taiwan, and perhaps as far north as Japan. Although they were not necessarily the original inhabitants of these areas, by virtue of their unique farming and fishing adaptation they occupied a coastal niche that had previously been unoccupied or underutilized.

By at least 5,000 years ago some of these coastal-adapted peoples began venturing farther eastward into the Pacific. Using improved canoes, probably dugout hulls stabilized by a single outrigger float, they sailed out of the sheltered waters of the Indonesian and Philippine archipelagoes and then along the north shore of the great island of New Guinea, where they settled on the small offshore islands and coastal stretches of the large ones. The question still remains as to exactly how these seafarers related to the less sea-oriented New Guineans whose ancestors had crossed over from Indonesia during the last glaciation when drastically lowered sea levels had decreased ocean gaps to the point where people possessing only rudimentary rafts could have made the voyage. Some archaeologists believe there must have been considerable mixing, both cultural and genetic, particularly in the islands of Bismarck Archipelago, and that people of mixed heritage from the Bismarcks were the ones who pushed farther into the Pacific.

Whatever the case, about 3,500 years ago, some adventurous canoe people from this sizable archipelago off the northeast coast of New Guinea began moving east along the chain of islands that was to lead them, after perhaps no more than a half-dozen generations, to the uninhabited archipelagoes of Fiji, Tonga, and Samoa. This quick thrust 1,800 nautical miles into the Pacific required a much more oceanic orientation than their ancestors had possessed, for the distances between islands grow from tens of miles to hundreds of miles the farther east you sail, and the rich flora and fauna of Southeast Asia fades away as you move from island to island. These seafarers responded to this oceanic challenge by developing an improved canoe—the double-canoe made by lashing two hulls together—to carry the colonists and their supplies on the lengthening voyages and by carefully transporting taro, yams, bananas, and the other plants they cultivated, as well as their pigs, dogs, and chickens, and then skillfully transplanting them on the fertile but biotically impoverished islands they found in the ocean.

Colonizing the large and lush archipelagoes of Fiji, Tonga, and Samoa seems to have occupied these seafarers and their descendants for many generations. As they settled into this midoceanic environment, the distinctive culture pattern from which all Polynesian cultures are descended developed from elements brought from the west. In the strictest sense then these archipelagoes, not some distant continental shore, form the long sought-after homeland of the Polynesians. Then, after a pause of many centuries, canoe prows were once again pointed toward the sunrise, initiating a second phase of rapid oceanic expansion. Sailing in large double-canoes, using periodic westerly wind reversals to move against the direction of the prevailing trade winds and currents, and navigating by naked-eye observations of the stars as well as cues furnished by the wind, swells, clouds, and birds, these voyagers—full-fledged Polynesians now—reached the Marquesas Islands far to the east by the first century before Christ. From this eastern outpost of Polynesian culture, and later from neighboring Tahiti, canoes sailed north, southeast, and southwest over thousands of miles of blue water to spacious Hawaii, tiny Easter Island, and massive New Zealand—points of a huge triangle almost as large as Europe and Asia combined. The discovery and settlement of these islands and every other inhabitable one within the Polynesian triangle, ranks as one of the outstanding achievements of the Stone Age.

In the late eighteenth century, just before massive European penetration into the Pacific, the Polynesians probably numbered upward of a million. A consensus among prehistorians holds that this population stemmed overwhelmingly from natural growth and not from the mass migrations of thousands of colonists sailing in great fleets of canoes. Given the time depth of thousands of years, the potential for growth from a small beginning is great. For example, the number of living descendants of a single couple who might have landed in Tonga in 1200 B.C. would have reached, at a compound growth rate of less than 0.5 percent per year, 1 million by A.D. 1800. Although no one seriously entertains the possibility that all Polynesians stemmed from

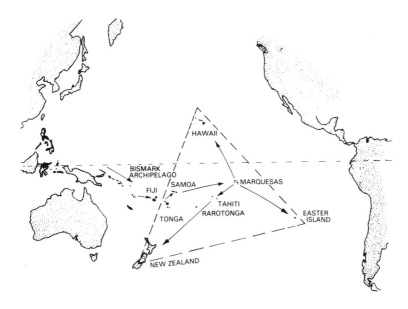

Fig. 10.1. Probable path of the Polynesian migration.

an oceanic Adam and Eve, the dominant opinion is that the number of people who actually migrated was comparatively small.

Just how many migrants may have been involved is impossible to specify. Traditional tales of colonization are typically obscure and at best disappointing as sources for facts and figures. Neither can archaeologists indicate definitely the size of initial colonizing parties nor the number of any additional migrants. Still, a judicious interpretation of legends and excavations would probably support the thesis that a majority of island populations were founded by migrants from single canoes and that the number of subsequent arrivals was probably few, except for those from neighboring islands of the same archipelago.

If we were to assume that a typical migrating party consisted of some 25 men, women, and children, and that the size of island communities from which they came and which they founded varied from 500 on small atolls to 50,000 on large high islands, the emigrants would constitute from 1 in 20 to 1 in 2000 of the source communities. Although such figure juggling is at best a speculative exercise, these ratios are certainly more realistic than assuming that whole communities or large fractions of them regularly uprooted themselves to sail to new islands.

But if this long process of Polynesian settlement involved only a trickle rather than a torrent of canoes, are we right in assuming that the migration was in large part purposeful? A generation ago a disarmingly simple thesis of Polynesian settlement gained wide credence: It all happened by accident. According to this view, because it was supposed that Polynesian canoes, navigation, and seamanship were not good enough for the task of undertaking planned voyages of migration over great distances, Polynesia could only have been settled through a long series of drift voyages in which generation after generation of hapless seafarers, pushed off course by wind or current while sailing between closely spaced islands or fleeing a war- or famine-ravaged homeland, were fortuitously cast up on enough successive shores to populate the entire region. Although this thesis had some value in countering wildly romantic notions left over from the nineteenth century of great fleets of canoes sailing over the Pacific at will, subsequent experimental voyages in which reconstructed canoes were sailed over thousands of miles of blue water guided solely by traditional, noninstrument navigational methods have shown that this premise of Polynesian nautical incompetence, and hence the exclusion of purpose from the migration, was wrong.

A more balanced view allows for some settlement by involuntary drift voyaging, but emphasizes that it was the technological complex of finely crafted voyaging canoes, noninstrument navigation techniques, and ways of transplanting plants and animals, not the vagaries of wind, current, and maritime misfortune, which enabled the Polynesians to find and settle virtually every island in the heart of the world's greatest ocean. However, any such thesis of purposeful migration begs the question of why people would leave familiar shores to strike out into the unknown sea and colonize islands increasingly bereft of useful plants and animals.

Population pressure has often been invoked as the prime driver in this oceanic movement. That pressure from ethnic Chinese pushing down from the north might have forced the distant continental ancestors of the Polynesians to leave the Asian mainland forms an intriguing hypothesis but one difficult to test archaeologically, given the time depth and watery setting. That population growth, particularly in conjunction with disasters wrought by drought, hurricanes, or tsunamis, might have repeatedly forced islanders to take to the sea in search of new land would seem a much more testable idea for archaeologists to tackle, especially since traditional accounts from the prehistoric period and reports from the early years of European contact tell of people being forced to flee because their island had become too crowded or because of sudden, disaster-caused famine.

I doubt, however, that population pressure alone can account for the totality of this extraordinary migration. Other peoples bordering the Pacific must surely have experienced population pressures at some time or another. Yet they did not take to the sea. And, if population pressure did push the Polynesians' ancestors out of South-east Asia, why is it that the great island of Borneo so near the presumed starting point remained so sparsely populated? Furthermore, once canoes started leaving the Bismarck Archipelago and then later the Fiji, Tonga, and Samoa region, the migrants seemed to have pushed ahead to discover and settle island after island at a rate that would seem faster than could be comfortably accounted for by any mechanistic model specifying that people do not leave an island until the population reaches carrying capacity. Something more than hunger must have motivated the Polynesians to sail so far and wide over the Pacific.

Factors internal to the structure of Polynesian society were certainly involved. For example, in the Polynesian inheritance system the firstborn son of a chief ideally gained all, leaving his younger brother without a title or really much to do—at home, that is. An ambitious younger brother of a chief had, in addition to fratricide, a most adventurous option—organize a voyaging expedition and hope to find and settle a new land where he, as the senior ranking male, would be Chief. Yet although this and possibly other social factors might have spurred many to leave their home islands, I think that the Polynesian migration must be seen above all as a manifestation of a uniquely oceanic, expansionary world view.

Distinguished biologists such as the late Theodosius Dobzhansky have not been shy about attributing to humans a basic urge to explore. The whole history of Hominidae has been one of expansion from an East African homeland over the globe and of developing technological means to spread into habitats for which we are not biologically adapted. Various peoples in successive epochs have taken the lead in this expansion, among them the Polynesians and their ancestors. During successive bursts lasting a few hundred years, punctuated by long pauses of a thousand or more years, these seafarers seem to have become intoxicated with the discovery of new lands, with using a voyaging technology they alone possessed to sail where no one had ever been before.

Once their attempts to cope with the rising sea levels of the Holocene committed them to sea, the first pioneers of this lineage of seafarers had good reasons to keep going. The continental mind set of their distant ancestors would have faded as successive generations pushed farther and farther east, to be eventually replaced by the more accurate view that the world was covered with water through which bits of land were scattered. They therefore knew that in pushing into the open ocean they were entering not a vast empty region but one teeming with islands. What is more, after leaving the Bismarck Archipelago and outdistancing their less sea-adapted rivals, they would have realized that before them lay an ocean of islands accessible to themselves alone. What more invitation did they need?

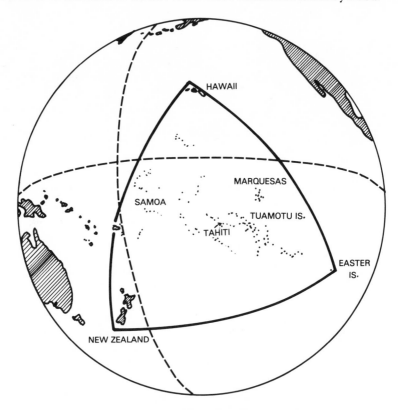

Fig. 10.2. The Polynesian Triangle—an ocean filled with islands.

Yet, at the time of first contact with European explorers, the Polynesian wanderlust had apparently been stilled. To be sure, as late as the early 1800s warlike Tongans were raiding neighboring archipelagoes, and periodically canoes from the drought- and conflict-ridden Marquesas Islands set out to sea in search of more inviting lands. But on most islands the people seem to have thoroughly settled down, content to focus their energies on earning a living from the land and on intrigues in political, religious, and other internal spheres.

Paradoxically, thinking of the Polynesians as having originally been a people uniquely adapted through their values and technology to finding and settling uninhabited oceanic islands helps to make this apparent cessation of their expansion more comprehensible. By the time the Polynesians had reached the Marquesas, Hawaii, and Easter Island at the eastern margin of the famous Polynesian triangle, they had run out of islands. Behind them lay populous Asia and the already inhabited islands through which their ancestors had sailed. Ahead, across thousands of miles of empty ocean, lay the linked American continents with a plethora of native groups, ranging from tiny bands of hunters and gatherers to the teeming populations of mighty empires. Once they had settled their triangle of ocean islands, the Polynesians really had nowhere to go.

We do not know how many canoes tried pushing farther east or if any made it to far-off continental shores. To sail all the way to the Americas would have been difficult but not impossible. For example, such a voyage might have been possible during El Niño periods when westerly winds and current reversals might push a canoe toward the Americas. Yet even if a canoe did manage to reach the American coast, survival would hardly have been easy for any colonists. As tropical, ocean-adapted voyagers used to colonizing virgin lands, any successful voyagers to the New World would have been ill equipped to cope with continental conditions and especially the numerous inhabitants upon whose territory they would be encroaching. Those Polynesians who might have reached the Americas would probably have been repulsed, killed, or absorbed, and thus would have left no more of an imprint than did their Viking counterparts on America's Atlantic shore.

What we do know is that the various island societies within Polynesia developed in a unique, if not total, degree of isolation from the rest of the world. To be sure, there were some contacts. Eastern Polynesians received the magnificent gift of the sweet potato, a South American cultigen, probably from some stray raft voyagers from that continent, although a Polynesian round-trip to South America and back cannot be ruled out as the mode of transport. Polynesians in the west had a more intense contact with the complex of peoples who had followed their wake and who virtually transformed Fiji from a Polynesian outpost into a Melanesian-Polynesian melting pot. Further, we cannot discount the possibility that a dismasted Japanese fishing boat or a disabled Chinese junk might have been carried by circular North Pacific currents to Hawaii or some point in Polynesia farther south. However, by and large Polynesia was on its own for thousands of years, its people able to follow their own destiny until voyagers from another ocean irrevocably broke their isolation.

What kind of people were these early seafarers? Trying to reconstruct the nature of their society is an even more hazardous task than attempting to trace their movements. Clues from archaeology and oral traditions are of some help, but ultimately we are forced to fall back upon a comparative analysis of the individual island societies that resulted from the migration. By factoring out what is common throughout Polynesia from what is unique to each island society—on the assumption that shared elements were part of the original pattern—we can model some of the main features of early Polynesian society.

Hereditary chiefs and a pervasive system of ranking defined by genealogical position were central to early Polynesian society. It was a society of structured inequality: Those most directly descended from the founding ancestors, and ultimately the gods, were the group's highest ranking citizens, and ideally the senior male among them would be the politicoreligious chief of the community.

Some anthropologists have proposed that ranked societies developed in Polynesia because of the need to control large island populations and organize their economic life. Although this may explain, as we shall see, the intensification of social stratification on large, populous islands, it does not tell us why ranking should be so pervasive throughout Polynesia, evident on even the smallest and most sparsely populated islands. Furthermore, this theory runs into trouble when we consider that many other populous Stone Age societies were not organized along principles of hereditary ranking. What is more likely is that social organization based on genealogical criteria was a basic feature from the time when small groups of seafarers first began to penetrate the Pacific.

Although we do not know exactly what led to the emergence of this chiefly way of organizing society, it certainly seems to have served these seafaring colonists well. Just as the ability to sail over the sea, developed originally in sheltered Southeast Asian waters, was preadaptive for the later expansion into the Pacific, so, I propose, was this chiefly system preadaptive to the task of exploring the sea and colonizing newfound lands. A chief whose lineage goes back to the gods can inspire his people; he can command their labor to build a voyaging canoe and prepare an expedition; and he can lead them over the seas toward faraway and possibly unknown lands. What is more, as those who have been to sea know, a colonizing expedition commanded by a leader of such acknowledged authority would have had a much better chance of successfully reaching a distant shore than one in which many voices constantly contended for authority. (Compare, for example, the exquisite voyages of Captain Cook with the near anarchy that plagued Magellan's divided command.) Similarly, the task of organizing life in a newly founded colony, particularly during the first years when the very survival of the community was so much at stake, must have benefited greatly from the unitary leadership provided by a hereditary chief, as well as the agreement among his junior kinsmen and followers as to who was in charge. For small groups made up largely if not exclusively of related families, this typically Polynesian allocation of authority along genealogical lines would have gone far toward providing the firm structure needed for overseas expansion.

That this Polynesian feature was not universal in societies at similar levels can be realized by glancing at the type of social structure dominating New Guinea. There inheritance counted for little or nothing. A man became a leader through his own accomplishments (usually in food and wealth production and exchange), and he maintained his leadership only so long as he continued to be successful in his economic endeavors. The resultant system of incessant jockeying for power as individual leaders rose and fell according to their fortunes would hardly seem ideal for mobilizing labor and for providing the continuous and universally accepted leadership needed for undertaking hazardous voyages of exploration and colonizing strange shores.

However, lest I leave the impression that these seafarers achieved colonizing success through social rigidity, I must emphasize that their society had built-in flexibility, which seems also to have been adaptive to the pioneering situation. For example, descent and inheritance was not reckoned purely patrilineally or matrilineally as in many tribal societies but in such a way that a person had some choice of whether to affiliate with or inherit from his father's or his mother's side. Furthermore, the children were easily and frequently adopted by kinsmen from either side of the family; the children themselves could seek succor and refuge from a wide range of kinsmen. In a pioneering situation, where the social fabric may have been frequently strained by high mortality at sea and during the first stages of the struggle to survive on an alien shore, as well as by chance imbalances in the sex ratio of children born to such small groups, such elements of flexibility undoubtedly served the cause of survival.

A legendary example of flexibility is contained in the traditional history of Manihiki-Rakahanga, an atoll community that was founded by a lone couple. As the traditional historians emphasized to the anthropologist who took down this tale, the ordinarily strong Polynesian prohibition against incest had to be broken for the nascent colony to survive beyond the first island-born generation. But what made for endless comment ever afterward was that a random skewing of the sex ratio in that first generation forced the founding father to do double duty. Because only daughters were born to the couple, he was forced to mate with them to produce the needed males.

The dispersion of such small, hierachically structured (yet flexible) groups over the ocean resulted in the creation of dozens of separate island societies with populations ranging from the hundreds on small atolls to the hundreds of thousands distributed among the high islands of the largest archipelagoes. A case could therefore be made that this Polynesian experience anticipates the time when small bands of humans may scatter through the Galaxy to found multitudinous colonies, large and small, on far planets, asteroids, comets, and artificial worlds. If so, we might ask what the Polynesian case can tell us of the human consequences of dispersion and the subsequent growth of isolated daughter colonies.

Although there is some evidence of microevolutionary changes in blood types, head form, and other physical characteristics, the populations of the dispersed island communities were apparently not separated long enough or were not sufficiently isolated from one another for major biological divergence to have developed. But the degree of separation and isolation was apparently sufficient to encourage the appearance of marked cultural differentiation. In many cases it is as if the successive generations of a society unconsciously seized upon one feature of the common Polynesian heritage for intense elaboration. For example, the Easter Islanders focused much energy on carving and erecting huge stone statues; the Tahitians became preoccupied with religious cults and rivalries; and the Marquesans specialized in tatooing to the point of covering the whole body with intricate blue-black tracings. Although ecological circumstance may have played a role in their beginnings, each of these cultural elaborations seems to have taken on a life of its own. The mutely staring statues scattered along the slopes of Easter Island are a case in point. Although the lack of good timber on this desolate island may have led sculptors to turn to stone in the first place, this circumstance would hardly have forced them to make such huge statues or to carve so many of them. This cultural analogue of genetic drift, working unchecked in the isolation of the windswept ocean, seems to have produced the most exaggerated of cultural efflorescences.

Cultural differentiation, however, is not the only type of change apparent in Polynesian societies. Running through their development is the tendency for social stratification to increase markedly as population grows. On the larger and hence more populous islands, social distance between high- and lowborn generally widened to the point where a class of hereditary and despotic chiefs emerged. For example, in the major archipelagoes of Tonga and Tahiti (including the other Society Islands), which each had populations in the 50,000 to 60,000 range, and Hawaii, where the population reached a quarter million, there developed complex and highly stratified social structures dominated by a hereditary caste of chiefs who exercised despotic control over the mass of lowborn commoners. This contrasts greatly with the more patriarchal and benevolent structures that prevailed on the smaller, less populous islands and presumably among the initial colonizing groups.

The seeds for this development of such a high degree of social stratification were contained in the original Polynesian ranking system that gave the pioneering explorers a ready-made structure for efficiently organizing their canoe crews and colonizing expeditions. So long as populations remained small, say no more than a thousand or so, hereditary chiefs could act benevolently as senior relatives whose task was to care for and guide their junior kinsmen. However, as populations grew, the genealogical gulf between senior and junior widened, face-to-face relations could no longer be maintained throughout the community, competition for scarce resources became rife, and conflict between competing groups and rival claimants to high status became endemic; complex, highly stratified, and often militarized chiefdoms resulted. A social structure admirably adapted for small, colonizing ventures came to be radically transformed as the colonies grew and prospered.

But not all these oceanic colonies prospered in the long run. Archaeologists have found the remains of long-extinct Polynesian colonies on rocky islets hundreds of miles north of the Hawaiian Islands, on lonely atolls straddling the equator, and on rugged Pitcairn Island of *Mutiny on the Bounty* fame. These small, isolated, ecologically marginal islands were hardly ideal for settlement—as those who attempted to live there found, for they either died out or fled in their canoes to find more inviting islands. Even on sizable and fertile islands the descendants of the colonists eventually had to reckon with limited resources. At best, that reckoning resulted in restrictions that make a mockery of the stereotype of free-and-easy Polynesian living. At worst, the crunch between burgeoning numbers and a limited and fragile resource base led to starvation, intertribal conflict, and wholesale environmental and societal degradation.

When a canoe landed on an uninhabited island, the first order of business was building an economic base for survival. Even on a large and fertile island this was not all that easy, for until the taro, bananas, and other food plants brought on the colonizing canoe began to bear, the colonists had to subsist on fish, birds and eggs, and what edible roots and berries they might find on these biotically impoverished islands. Once they had made it through this initial period, however, the colonists and their immediate descendants had an entire island of valleys and mountain slopes for farming, as well as a range of reefs, lagoons, and offshore waters for exploitation. Polynesian pioneers responded to this opportunity just as other species do when coloniz-ing an empty island or, more accurately, an empty niche in an island ecosystem. They multiplied rapidly, quickly filling the new land.

Once an island became crowded and pressure upon resources evident, the Polynesians, like other island-colonizing species, were forced to change their strategy in terms of both production and reproduction. Extensive agricultural practices (such as shifting agriculture, in which successive sections of mountain slopes were cleared, burned, and cultivated for a few years and then abandoned) gave way to more permanent field systems and, where possible, to intensive irrigation systems—all in an effort to obtain more food from the limited land available. At the same time the islanders tried to limit population growth by such means as coitus interruptus, abortion, and infanticide.

Even on islands where some balance between population and resources appears to have been achieved, the human costs of coping with scarcity were all too evident. On Tahiti, for example, the 35,000 Polynesians living there at the time of European discovery were divided between high-status persons with full access to food and other resources, and low-status persons with limited access; there is some evidence that female infanticide was so widespread that a marked skewing of the sex ratio had resulted; and many Tahitians belonged to a religious cult that was dedicated to licentious pleasure and that forbade its members to have children. Elsewhere in Polynesia were islands where more Malthusian checks to population growth came into play: famine, war, and the forced expulsion of whole segments of the population.

Climatic disasters, such as typhoons and severe droughts, must have played a role in some of these cases. For example, periodic droughts in the Marquesas Islands led to famine, which set valley against valley in bitter conflicts that often resulted in the slaughter, expulsion, or flight of the weaker. However, although the return of the rains appear to have restored a temporary balance to the Marquesas, on some islands more catastrophic and longer-lasting effects of population growth and resource depletion occurred.

Despite their lack of metal tools and other forms of high technology, the Polynesians were fully able to degrade an island environment. Reef and lagoon life and shore-dwelling birds unused to predators were the first to feel the impact of these hungry, fecund newcomers. For example, in Hawaii and New Zealand colonists wiped out species after species of vulnerable birds. Then as they applied themselves to farming, the stone adze and especially fire took their toll. On some islands the resultant deforestation and erosion may have been beneficial in a "terraforming" sense: Soil washed down denuded mountain slopes filled in swampy valley bottoms and shore lands to increase the amount of land available for taro irrigation and other intensive agricultural practices. However, on many islands, especially environmentally marginal ones, the effect of uncontrolled agricultural expansion was hardly so benign. Again, that most unusual of Polynesian islands, Easter Island, furnishes the example.

This easternmost outpost, lying closer to South America than to most other Polynesian isles, was settled around A.D. 500. As the colonists multiplied (to as many as 10,000, says one authority), they cut back the forest to clear land for farming and to obtain timber for canoes, houses, and firewood and also for sleds and scaffolding to move and erect their massive stone statues. By the time the first European voyagers arrived in the eighteenth century, the island and its people had suffered a catastrophe. The forests were gone, replaced by windswept grasslands. Food, by then grown mostly in sheltered pits or behind stone windbreaks, was scarce; timber virtually nonexistent. The population, cut back to a few thousand, was dispirited and divided. They had quit making their huge stone statues and had toppled over those that once stood on the long stone altars of their great temples. What is more, they no longer could sail far out to sea, for their largest canoes measured hardly 3 meters in length: leaky, patchwork vessels made up of small scraps of wood painstakingly fitted and lashed together. The survivors had become trapped by a disaster of their own making, unable to regain their former prosperity and numbers or to recreate their ancestral voyaging canoes so that they might flee their lonely and degraded island.

After rather confidently sketching some 10,000 years of Polynesian and proto-Polynesian history and sociology, let me offer some caveats. The dispersion pattern within the Polynesian triangle is fairly well known, and recent archaeological work has enabled us to backtrack, almost from island to island, along the migration trail as far as the islands offshore New Guinea. Although the available evidence points from there to Southeast Asia as the ultimate source of the migration, we cannot yet draw firm arrows on the map for this earliest part of the trail. The reconstruction of the number of people involved, their motivation, their social and cultural institutions, and the changes wrought in these is much less sure than the bare-bones outline of their movements. Furthermore, although the resultant portrait of small bands developing a technology to explore the sea, of their founding colonies on the islands they discovered, and of the growth and transformation of these colonies might seem to provide a coherent preview from Earth history of what could happen in space, we obviously have no way of knowing whether that will be the case.

Given these problems, what then is the value of the exercise? I would claim that despite the imperfections of the record and the obvious differences in technology and scale, examining this extended example of migration on Earth is useful because it makes us think about the basic human issues of settling new worlds, issues that are likely to be otherwise ignored—until they are upon us.

The colonization of space will mean the scattering of men, women, and children among increasingly distant worlds where they will breed, live out their years, and die. It will involve complete social organizations, some stripped down and austerely designed to pioneer new worlds, others grown rich and complicated as prosperous, transplanted colonies. It will see the birth of many new civilizations from small and often skewed cultural samples; it will see the death of others. Reflecting on the Polynesian experience sensitizes us to the human issues raised by the prospect of space colonization. It forces us to think about the mix of motivation and technology that drives people to explore and settle distant, uninhabited worlds; to be aware of the need to design small yet complete societies adaptive for long voyages and difficult pioneering tasks; to consider how these austere societies will and must change upon the successful implantation and growth of a colony; to ponder the possibilities for rapid cultural divergence as each scattered colony assumes an identity and historical course of its own; and, finally, to anticipate that hardship and disaster await those groups who settle on worlds too poor to sustain permanent colonies or who carelessly exhaust the resources of more viable worlds.

But to go beyond the particular issues to the general significance of the Polynesian case, let me end this essay with a personal hunch. This odyssey of Stone Age voyagers may well be the event in our species' history most relevant to the projected expansion into space. It suggests that it may not be Earth dwellers who will populate space, but rather those space pioneers who, through generations of residence on the Moon, Mars, asteroids, or in space colonies, have thoroughly adapted to what we call the alien environment of space but that they will call home. Employing new technologies, new forms of social organization, these space-adapted people will be the Polynesians of humanity's odyssey among the stars.

References

Bellwood, Peter. 1979. *Man's Conquest of the Pacific*. Oxford University Press, Oxford.

Buck, Peter H. 1959. *Vikings of the Pacific*. University of Chicago Press, Chicago.

Dobzhansky, Theodosius. 1962. *Mankind Evolving*. Yale University Press, New Haven.

Finney, B. R. 1977. Voyaging Canoes and the Settlement of Polynesia. *Science* 196, 4296: 1277-1285.

Finney, B. R. 1979. *Hokule'a: The Way to Tahiti*. Dodd, Mead, New York.

Goldman, Irvine. 1970. *Ancient Polynesian Society*. University of Chicago Press, Chicago.

Howells, William. 1973. *The Pacific Islanders*. A. H. and A. W. Read, Auckland.

Jennings, Jesse, ed. 1979. *The Prehistory of Polynesia*. Harvard University Press, Cambridge.

Kirch, Patrick V. 1980. Polynesian Prehistory: Cultural Adaptation in Island Ecosystems. *American Scientist* 68, 1: 39-48.

Macarthur, Robert H., and Edward O. Wilson. 1967. *The Theory of Island Biogeography*. Princeton University Press, Princeton.

McCoy, P. C. 1976. *Easter Island Settlement Patterns in the Late Prehistoric and Protohistoric Periods*. Easter Island Committee, International Fund for Monuments Inc., bull. 5.

*11

Richard B. Lee

MODELS OF HUMAN COLONIZATION: !KUNG SAN, GREEKS, AND VIKINGS

Like most of the social scientists at our meeting, I was treated to a veritable feast of technological speculation and extrapolation on the questions of interstellar colonization. The organizers of the conference on Interstellar Migration have encouraged the social scientists to rein in the space scientists and ground them in the social and historical realities. Taking them at their word, I will focus on two topics commonly left out of the space scientists' equations: human history is one, the other is human behavior. To paraphrase Santayana, if we do not look seriously into our past and present, history dooms us to repeat our mistakes in the future.

Let us consider examples of past human colonization and try to draw what lessons we can to seek precedents for the problems we will face in colonizing space. I want to look at three kinds of societies: a hunting and gathering people, drawing on my own work with the !Kung San of the Kalahari Desert, and two mercantile peoples, the ancient Greeks and their colonization of the Mediterranean, and the Vikings and their movement into the North Atlantic.

How have human beings reacted in the past to the problems of occupying new territories? There are a number of questions we can ask. First of all, we are interested in the causes of colonization. Is it overpopulation? the desire for plunder? the desire for trade? Are there problems of inheritance laws, for example, that create movements outward? Second, we are interested in the course of colonization from the first reconnaissances or explorations to the planning and execution of major movements of peoples. We are concerned with the means of transportation, the kinds of vessels or land vehicles used, the size of the groups involved, the organization on shipboard, the organization of the new colony, and so on. We can also ask questions about the relation of colony to mother country. Is the colony subject to, independent of, or dominant over the mother country? History gives us examples of each of these. We are interested in relations between colony and colony: warlike or peaceful? We are interested in the question of the destiny of colonies: Do they outgrow the mother country? Do they, alternatively, lose contact with the mother country and die out? There are a range of possibilities between these two extremes. We are also interested in questions of time frame, sequencing, and geography: the rate of spread of the colonizing movement, the shape of the colonizing wave front. We will not be able to cover all these questions for all three case studies, but we will do well to keep them in mind.

Kung Foragers

My first case is not an example of colonization at all. The !Kung San, formerly called the Bushmen, live in Botswana and Namibia (fig. 11.1). Traditionally, they lived by hunting and gathering, the way of life that was, until 10,000 years ago, the universal mode of human existence. The total population of the !Kung today is about 15,000, most of them now living in nonhunting and gathering situations on Herero Tswana cattleposts or European farms. Only about 10 percent continued to live as hunters and gatherers into the 1960s. My work in the Dobe area of Botswana starting in 1963 has totaled about three years of field research.

The Dobe area !Kung are a stable population of about 460, remarkably close to Birdsell's notion of a stable breeding population of 500. They are dispersed at nine water holes and divided into twenty-five living groups with a mean size of eighteen to twenty people. Subsistence through the 1960s was by hunting and gathering. Given the popular conception of hunting peoples living a life of hardship, one of the surprising things about the !Kung is that they have an

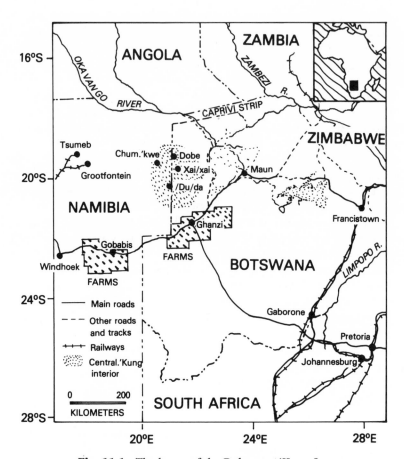

Fig. 11.1. The home of the Dobe area !Kung San.

abundant food supply and a lot of leisure time. The main food crop is the wild mongongo nut, millions of which are harvested every year.

Social organization is simple. The main unit is a camp, flexible in composition, around which social life revolves. Females have a rather late menarche—girls don't reach puberty until 15 or 16 and their demography is characterized by very low fertility and high infant mortality. Nancy Howell, in a thorough demographic study, found that mean completed family size was 4.6 and that about 45 percent of those born died before reaching maturity. Low fertility and high mortality yield a very low but still positive growth rate. Howell also found a very long adult life expectancy. If one survived childhood, one could expect to live well into the upper years. About 10 percent of the population during my fieldwork was more than age 60, a proportion similar to that of elderly in the U.S. population around the turn of the century.

Two key points need to be underlined about the !Kung. First, they are an egalitarian people, with no chiefs or headmen. This rather attractive political system has its advantages and disadvantages. One disadvantage is that there is no central authority for resolving disputes, and they do not have the option that we do of going to court or to the police in the event of a dispute. They have to resolve everything themselves. One advantage is that the core of their adaptation is based on sharing: the giving of food in a familial way without the immediate expectation of return. The !Kung and many other hunter-gatherers carry the sharing principle farthest of any human society. They have a very low development of private property, and it seems that their long-term survival has worked very well based on this principle of sharing.

It is difficult to speak of colonization with reference to societies like the !Kung. Band societies, when they migrate, go through a sort of a hiving-off process; a kind of Brownian motion is involved in their dispersal mechanisms. It is not systematic migration at all. Yet it is worth noting that more of Earth's surface has been peopled by these Brownian movements than by the systematic migrations of later times. When the Europeans began their sea-borne expansion, they found the rest of the world already occupied.

The Greeks and Vikings, by contrast, were organized into chiefdoms and rudimentary states. Colonization here was relatively systematic and often centrally organized. Further, water is a medium similar in some respects to space. Being a foreign medium to air breathers, it requires a certain level of technology to cross effectively and has many of the dangers that space will present to future colonists. In Greek and Viking colonization small groups set out in fragile craft, a situation that would seem to require a strong organization focused around a founder, a central leader. These people, I think we can say, were motivated by a spirit of adventure; they constituted a skewed sample of the population at large, that is, people who were willing to give up their comparative comforts at home and put up with hardship, isolation, and loneliness. The migration of the Greeks and Vikings were high-risk enterprises. The rewards could be great, but some of the colonies failed, others were lost at sea, and others lost contact with the mother country.

Greek Colonization of the Mediterranean

The preclassical Greeks were a textbook example of massive and rapid colonial expansion. The period that interests us most is known in Greek history as the Archaic or Geometric period, 800 to 500 B.C., which was marked by rapid social development after a period of "dark ages." The expansion was triggered around 800 B.C. by three factors: the rediscovery of literacy with the new Phoenician alphabet (the forerunner of the modern alphabet), the diffusion and rapid coming into use of iron technology, and the rapid growth of production and population. The outcomes of this explosive growth of productive forces were several. First, in the home cities of the Greeks, intense social inequality rapidly developed. Peasants were forced off the land and sold into slavery. Second, there was the development of one-man rule, the tyrannies that were so popular in Greece from 700 to 500 B.C. The third outcome was colonization. There was a massive out-migration from the Greek core area, and it clearly must be seen as a response to population pressure. I liked the distinction made earlier in our discussions between two kinds of responses to population pressure: Are we talking about peoples who are fleeing it or relieving it? I think in the Greek case the volume and size of the migration indicate that both aspects were in force. There were some cases where cities actually exported a significant proportion of their population.

The Greek colonization wave is traditionally dated as beginning around 750 B.C. The earliest moves were to Sicily (perhaps the world's oldest colony) and to the northern Aegean. In the early period colonies were usually offshoots of the mother city, which issued foundation decrees (some have been preserved) and which designated a leading citizen around whom the colony revolved. The actual ships employed were simple galleys with one or two rows of oars, and the distances were not long—not more than a two-day maximum between landfalls. The Mediterranean is very much like a pond compared with the central Pacific.

In the first two hundred years of colonization, more than 180 colonies were established. Most colonies were usually city-states, modeled on, but independent of, the mother city. There were exceptions. Corinth, for example, preferred direct control and developed quite a little empire with eight colonies under its control. Trade was clearly a factor in the founding of colonies. From the western regions of the Mediterranean, Greece obtained tin, silver, and gold; from the Black Sea, gold and grain, especially the latter. Much of the food that fed Athens in the Golden Age was grain from what is now the Ukraine.

Unlike the Polynesian islands, all of these lands were populated, and Greeks got along with the natives surprisingly well. In Sicily there were some wars and in some cases the colonists displaced the natives. But more often than not, relations between Greeks and natives were amicable. The native people benefited, it seems, from the Greek presence and trade; they received highly desired trade goods—pottery, wine, and olive oil—as well as exposure to Greek culture. In fact, it is clear that there was more fighting among Greeks than between Greeks and natives. For example, in 660 there was a war between Corinth and one of its main colonies, Korsyra, the modern Corfu. Some of the 180 colonies can be seen in figure 11.2. They stretched from Spain in the west to Egypt in the east and from the Crimea in the north to Libya in the south.

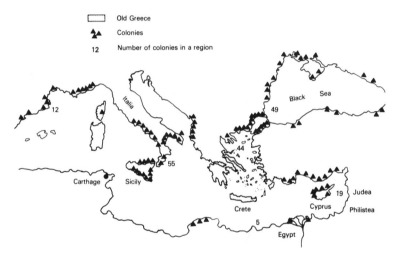

Fig. 11.2. There were 184 Greek colonies established between 750-500 B.C.

Some aspects of the Greek colonization movement strike chords for us and evoke modern and future colonization movements. Colonies tended to be richer than the mother cities. For example, the town of Sybaris, on the toe of Italy founded around 708, was fabulously wealthy. The word *sybaritic*, meaning "rolling in luxury," is derived from this colony. The Sybarites drew wealth from a relatively rich hinterland and animated a whole economic region both through trade and production.

The image of the future in which more humans will live on space colonies than will live on Earth is something that has precedent in the

Greek experience. By 500 B.C. there were more Greeks living outside Greece than there were in nuclear Greece itself. Yet—and this is something we should reflect on too—when we look at the history of Greece, we see that most of the historical material comes from nuclear Greece. Aristotle was based in Athens and not in Sybaris or in Sicily, although it is true that some of the famous philosophers—such as Pythagoras—did work in Sicily and Italy. Philosophy, science, and politics all flourished in the core area, and in that respect the colonies never quite lost their status as rural peripheries.

The Greek data present some interesting perspectives on the question of sequencing. One theme of the interstellar migration conference has been the building of models of an expanding wave front of colonization. I assume such models are based on animal ecology dispersion models as much as anything else. In the Greek data there is a very poor correlation between geographic distance from the mother country and the date of settlement. Distant points were settled first, and later colonies could be much closer in. The Sicilian and southern Italian regions give us some good examples (fig. 11.3). We see that, for example, Syracuse on the east side of Sicily was founded in 734; three other colonies on the same coast were founded about the same time. However, contemporary with these are very early colonies as far north along the western shore of Italy proper as Cumai on the Bay of Naples. Conversely, fairly close in to the nuclear area of Sicily are colonies founded at very late dates such as 598. So I cannot construct a reasonable set of isomorphic lines that would create a wave front model of colonization of the kind we have been looking for. (I should mention that there is another problem, a caveat about these founding dates: They come from ancient sources and they don't always square with archaeology.)

The point is that the founding of Greek colonies was structured by many factors other than mere distance: the presence of good harbors, the availability of suitable defensive sites, whether the natives were friendly—just to mention a few. Similarly, I would argue, a multiplicity of factors other than distance will shape the choice of colonizing sites in space.

Another complicating factor is that, as we might predict, colonies started founding other colonies almost immediately. For example, only five years after Naxos was founded, it in turn founded a secondary colony named Leotini, and fifty years later primary colonies were still being established elsewhere in Sicily. I cannot see any simple equation that would handle this in a way that would conform to wave front colonization.

Fig. 11.3. The numbers accompanying the names of these Greek colonies in Italy and Sicily are the founding dates. Those that are underlined indicate second generation sites—colonies of colonies.

Vikings in the Atlantic

Let us now look at the Vikings. Predatory as the Greeks were, the Vikings were much more predatory. The Vikings' career of colonization begins with a violent eruption on the scene around A.D. 800 and involves a great deal of fighting; even the names, like Eric Blood-Ax, conjure up a certain image. The extent of their spread was spectacular. By A.D. 1000 the Vikings stretched from North America to Constantinople. Scandinavian peoples moved both east and west and three divisions can be distinguished, with a great deal of mixing (fig. 11.4).

The Swedes mainly went east and founded the Varangians, who in turn provided the ruling dynasties of what later became Russia. Their great trading cities were Novgorod and Kiev, and they extended their

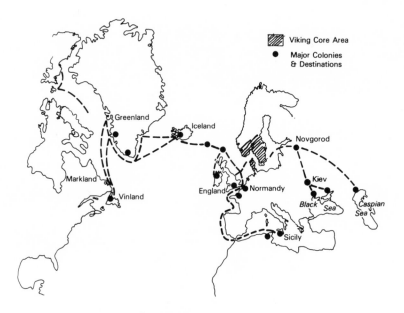

Fig. 11.4. Viking colonies east and west.

trading posts to the Black Sea and down the Volga as far as the Caspian Sea. Some of the best material we have on this period is from the Arab chroniclers who encountered the Varangians at the mouth of the Volga.

The Danes went west and fell on England with fury in 793, sacking the monastery at Lindesfarne on the northeast English coast. From there they proceeded to occupy the northern two-thirds of England in an area that came to be called the Danelaw. Later, however, the Danes were defeated by the more indigenous English kings of Mercia, after which they fade from the scene.

The Norsemen proper, who were based in southwestern Norway, had the greatest impact on Western Europe. Initially, they went west to Ireland and Scotland and south to France. They founded Normandy, getting in on the ground floor as the ruling class of high feudalism, and later, of course, conquered England in 1066. They also set up a Norman kingdom in Sicily and North Africa. But the migration that interests us most is the oceanic island hopping that led them to North America.

What set the Vikings in motion first of all was clearly overpopulation. The arable areas of southwest Norway were limited (as was the Greek homeland) and could not sustain a large population growth. There is some evidence that the Vikings overseas and at home had an extremely high birthrate. The Vikings were polygamous and leading men had ten or more wives and concubines. The Arabs, who were no slouches themselves, devote page after page in their chronicles to the sexual proclivities of the Vikings. They observed the frequent public performances of multiple sexual intercourse on the occasions of funerals and marriages. Highly visible sexuality doesn't necessarily mean a correspondingly high birthrate, but it is suggestive.

The question of whether it was a climatic optimum that facilitated the exploration and settlement of the North Atlantic is a topic that needs investigation. The latter half of the story definitely has to do with a climatic deterioration: The Little Ice Age from A.D. 1200 to 1600 is implicated in the abandonment of settlements in Greenland and elsewhere.

A second factor that set the Vikings in motion was inheritance laws. With a system of primogeniture, first sons inherited all and younger sons were basically out of luck. A similar factor has been noted for Polynesian migrations.

A third factor was political rivalry. Norway was ruled by petty kings at home, and much of the outward movement as described in the sagas was initiated by leading men, chiefs, thwarted in some kind of political maneuver at home, who got on board ship and with their followers set out to look for living space.

The ships were beautiful, magnificent craft of 60 to 90 feet long with prows carved in beastly shapes. They were propelled with a single row of oars and sail power. The keels were built to be flexible, which made them more seaworthy, allowing the ships to absorb the shocks of the waves. They had very shallow draughts and thus could travel upriver long distances. At freeze-up they were hauled out and beached for the winter. Crews numbered around twenty-five, with twelve or so rowers on a side, and there was room for passengers and livestock. The navigators used latitude sailing. By keeping the polestar at a constant height above the horizon, they could cross large spaces of open water between landfalls. They used winds, of course, and bird lore as well. One Icelandic saga mentions a man called Ravenflokli, who around 860 took three birds with him on a trip west. When he released the first raven, it immediately flew back to their starting point. The second bird he released flew up into the air and landed right back on the ship. A few days later the third bird was released and flew straight west and led him to a land which he named "Iceland."

Let us look at the sequencing of the Viking move west. They invaded England in 793; they colonized the Shetland and Orkney islands around 800 and the Faroe Islands in 825. An interesting historical footnote is that on all these remote islands the Vikings found Irish hermit monks in residence. The monks had discovered these islands long before the Vikings and were there to forsake all human contact and to commune with God. With the Vikings' arrival, the neighborhood went downhill. Then in 860 the Vikings reached Iceland, and by 870 they had begun to settle there. Iceland is a good example of a very rapid filling up process; by 930, after only sixty years of systematic settlement, the land was extensively settled. Some scholars estimate that by then the population may have reached as many as 30,000 to 60,000, impressive numbers particularly since now, after centuries of development, the population is only about 200,000. Whatever the exact numbers, some Icelanders certainly felt crowded and by 930 had begun to think about moving on.

Iceland had forests and excellent grazing for sheep; then there were the wonderful resources of the sea. Unlike Greek colonization, where mostly males went out and married local women, Norse colonization was not an all-male operation. In fact, the Norse trait of migrating in family groups may have contributed significantly to rapid population growth. But population growth was not all natural increase, of course: There was constant traffic between Iceland and Norway and between Iceland and Ireland. The Celts formed part of the slave population; Celtic contribution to the contemporary Icelandic gene pool is estimated at about one-seventh.

At any rate, in 982 an interesting character from our history books named Eirik the Red was banished for homicide from Iceland, after having previously been banished from Norway for the same offense. Unable to go east, he decided to use his three years of banishment usefully and so set off to the west. He sailed to explore Greenland whose sighting had recently been reported. When he returned to Iceland with exciting tales of a vast land teeming with game, settlement fever overtook the Icelanders, and twenty-five boats sailed off in 986, of which fourteen arrived safely in Greenland, bringing about 450 colonists. They set up an initially very successful settlement in Greenland. In the same year, 986, a man named Bjarni Herjolfsson chased the Greenland expedition—he was too late to go off with it—but missed Greenland entirely, and after many days' sailing reached Labrador. Interestingly enough, the sagas relate that he never landed; he just coasted up and down the shore of Labrador, crossed to Baffinland, and by dead reckoning got himself back to Greenland. It was based on his information that Leif Eiriksson set out in 1000 for North America. Leif used Bjarni Herjolfsson's ship and some of his crew. Retracing the same route, he reached North America with landfalls on Baffin Island, then Labrador, and finally Vinland.

There has been much debate about the Vinland story. Clearly, the first settlement was in northern Newfoundland. Helge and Anne Ingstad have been excavating an apparent Norse site at l'Anse aux Meadows at the northern tip of Newfoundland, which is probably the Vinland of Leif Eiriksson.

There were subsequent trips: about four voyages in all during the next decade. A man named Thorfinn Karlsefni married Eirik the Red's daughter and went out with three boats and about 160 men, women and children, and livestock, with the intention of making a permanent settlement in North America. It is not clear where the settlement was situated, but they lasted only three winters before being driven off by the unfriendliness of the climate and the native population. On the plus side, the sagas relate fabulous trading deals, in which the natives give Thorfinn a valuable fur for a cup of cow's milk or several furs for a strip of red cloth. But in spite of these bargains, the colony failed and returned to Greenland. During the next three centuries, there were several other voyages to North America, some voluntary and some involuntary, but it was not until the sixteenth century that the next permanent settlement of Europeans in North America was attempted.

Let us go back to the Greenland settlement, for it offers us some insights. It grew to a maximum size of about 3,000 people, with a dispersed settlement pattern. There was an eastern and western settlement, both of them on the southwest coast of Greenland. After A.D. 1200 ecological and political changes conspired to reduce seriously the colony's chances of survival. The period 900 to 1200 has been called the Little Climatic Optimum, a period of warmer temperatures and receding ice; most of the westward activity was accomplished in the window provided by this climatic change. In the period 1200 to 1600 there were colder temperatures, shorter growing seasons, and advancing ice. The routes of travel between Iceland and Greenland, for example, had to be drastically altered because the old routes were blocked by ice. The Greenlanders had advanced up the coast as far as 79° north latitude, as attested by runic inscriptions on stones found at that latitude, both on the Canadian arctic islands and on the Greenland coast.

But by 1350 the northern of the two major Greenland settlements had to be abandoned; by 1380 Greenland's sole remaining settlement lost regular contact with Iceland and Europe. Very infrequent and intermittent contacts continued every thirty years or so; there was one voyage in 1385, another in 1406, another in 1448, and the last, in 1476, by English merchants from Bristol. Then silence; the colony died out.

There are many theories to account for the dying out of the Greenland colony. One is that the colonists were killed off by the Eskimo. Another theory, citing skeletal evidence, argues that they physically degenerated and died out. Still another argues that they blended with the Eskimo, that is, "went native," and it is probably true that some of them did. This view gains support from a 1637 Vatican document based on an earlier text from 1342, which states: "The inhabitants of Greenland of their own will [have] abandoned the true faith and the Christian religion . . . . and went over to the people of America." Although the Eskimos were in Greenland before A.D. 1000, they were not in southwestern Greenland when the Norse first arrived; they only began moving south after 1200 when the growing cold conditions forced them to move down the coast.

There are some more extravagant theories of the fate of the Greenlanders as well. One fanciful idea is that they made a last desperate escape attempt to America and disappeared into the American wilderness. This is sometimes connected with the supposed findings of Norse relics in such inland locales as northern Ontario and Minnesota.

All these scenarios might have some truth, but the mystery remains. It is a cautionary tale for us: Not everyone wins at the migration game. Colonies can be lost, and Greenland is a well-documented example. A footnote to the story is that Greenland was reoccupied by Danish settlers after 1700 and today has a population of 50,000, who are Scandinavian-Eskimo hybrids, with their own parliament, newspapers, and a thriving culture.

Space Colonization: Sybaris or Greenland?

Before leaving the matter of human colonization, I would like to report some observations on a recently discovered culture—that of the space scientists who were at the Interstellar Migration Conference.

As an anthropologist studying this culture, I find it at least as exotic as the cultures of the !Kung, the Greeks, and the Vikings; in some ways more so. Their view of space gives me an interesting insight into their view of the world. They seem to have a tremendous commitment to and faith in science and technology and their ability to solve our problems. I have the feeling that their visions of space are really a projection of the American free enterprise system into the cosmos. For example, something about these wonderful colonizing wave models stuck in my mind. In a way it is the American utopia and manifest destiny all over again: an unlimited frontier, vast resources—like winning the West with no Indians. Gerard O'Neill's *The High Frontier* makes this symbolism explicit.

Implicit in the ethos of these space scientists is an interesting view of women contained in the models of rapid population growth. The wave front model, for example, appears to assume a doubling of the population every generation or so. How is this doubling to be achieved? Are they really casting women in the role of reproductive machines? We have focused almost exclusively on the geometric progression of numbers involved in founding colonies and have largely ignored the human implications of such numbers. Some of them may argue that no, we are not casting women in that role at all, that everything will be done in a test tube. But that seems to entail an equally grim view of our future. If the future in space is a choice between women as baby-making machines or machines as baby-making machines, you will have to count me out on both scenarios.

The upshot of all this is to inject a somber note in an otherwise optimistic proceeding. We need to learn a little humility and at least to become aware of our own unexamined assumptions. These scientists aspire to the stars and yet their vision is profoundly limited by the blinds of one culture at one point in history. History is messy, and the human material we are working with is messy. Let us at least try to be aware of the triumphalism that I hear again and again: "We're going to space, it's our destiny." Such sloganeering strikes a hollow note for those of us who are far from sure that technology will solve all of our problems.

There are two big unknowns in this whole business of colonizing space, and we have looked at only one. The first, of course, is the extraterrestrial unknown: What is out there? But the second big unknown is the search for terrestrial intelligence: What is down here? If we are as smart as we claim to be and can go to the stars, why can't we use our considerable intelligence to solve the problems on Earth before we export them to space?

A final word: Why bring the !Kung into this whole discussion at all? They are not an example of colonization. They have been out in the arid interior of southern Africa a long time, and there is no evidence that they have migrated from elsewhere. The !Kung do tell us three important things that we would do well to reflect on. They tell us, first, how to live in a small group. Second, they tell us how to be self-sufficient, and third, how to be able to do this for a very long time. One of the most exciting things about this whole conference for me is the vision of social engineering. I will leave to the rocket men their equations; but what the social scientists could contribute is the designing of a human experiment: a closed-system, multigenerational human experiment. Imagine twenty groups of twenty-five spacefarers tethered to an interstellar sail, traveling through space, and getting together at a common location once or twice a year for a thousand years! When we are planning a space colony or a space ark to travel to another star, we may well have to rediscover the lessons that people like the !Kung can teach us.

References

Bickerman, E. J. 1980. *Chronology of the Ancient World.* Thames and Hudson, London.
Bryson, Reed A. 1977. *Climates of Hunger: Mankind and the World's Changing Weather.* University of Wisconsin Press, Madison.
Howell, Nancy. 1979. *Demography of the Dobe !Kung.* Academic Press, New York.
Ingstad, Helge. 1969. *Westward to Vinland.* St. Martin's Press, New York.
Jones, Gwyn. 1968. *A History of the Vikings.* Oxford University Press, New York.
Lee, Richard B. 1979. *The !Kung San: Men, Women and Work in a Foraging Society.* Cambridge University Press, Cambridge and New York.
Myres, J. L. 1925. The Colonial Expansion of Greece, In J. B. Bury et al. *The Cambridge Ancient History*, vol. 3. Cambridge University Press, Cambridge. Pp. 631-690.
Newby, Eric. 1975. *The World Atlas of Exploration.* Mitchell Beazley, London.

"The maps of the census reports show an uneven advance of the farmer's frontier, with tongues of settlement pushed forward and with indentations of wilderness. In part this was due to Indian resistance, in part to the location of river valleys and passes, in part to the unequal forces of the centers of frontier attraction. Among the important centers of attraction may be mentioned the following: fertile and favorably situated soils, salt springs, mines, and army posts."
—*Frederick Jackson Turner*

"We cannot launch our planetary probes from a springboard of poverty, discrimination, or unrest; but neither can we wait until each and every terrestrial problem has been solved. Such logic two hundred years ago would have prevented expansion westward past the Appalachian Mountains, for assuredly, the Eastern seaboard was beset by problems of great urgency then, as it is today."
—*Michael Collins*

✳12

Ben R. Finney

THE PRINCE AND THE EUNUCH

Late one June afternoon I was traveling by train from Lisbon to Lagos, a small port at the southwestern corner of Portugal. As the train slowly wound through the hilly, sun-baked country, I fell to talking with the Portuguese gentleman sitting next to me, a banker from a small town near Lagos. When he asked why I was going to Lagos, I replied, "To go from there to Sagres, to see where Prince Henry once lived." Fortunately, he understood my halting Spanish, but before I could explain why I was going to Sagres he launched into a tribute to the historical importance of that barren point that juts so defiantly into the Atlantic. "Sagres," he announced, "was once as important as your Cape Kennedy, and Prince Henry was like your astronauts." He went on to explain, without my prompting, that the exploration of the sea inaugurated by Prince Henry was in its time as important as exploring space today.

I was astonished. For months I had been examining the analogy between the European Age of Exploration and the current Space Age, and was just then on my way to Sagres to make a pilgrimage to that Cape Kennedy of an earlier era. Yet I probably should not have been so surprised. The Portuguese are so obviously proud of their navigating prince and of the role Portugal once played as the world's leading exploring nation; and the analogy is widely made, though mostly, I had thought, on my side of the Atlantic.

American space enthusiasts are particularly fond of pointing to European maritime expansion, especially the discovery of the New World and its subsequent colonization and development by Europeans and their American descendants, as a precedent for what should happen in space. And, in this day of budget cuts and lagging official support, many admire the vision and persistence of explorers like Columbus, as well as the astuteness and daring of backers like Isabella, and openly wish that we had such farsighted leaders today.

Yet the applicability of this European expansion model comes into question when we recall that the explorers discovered lands already populated and that the colonization of those lands entailed the exploitation, enslavement, or death of millions. Should we be the only intelligent species in the Galaxy, or at least the only one in our neighborhood, this model would not apply. Should we be one of many intelligent life forms sown throughout the Galaxy, the prospect of colonizing already inhabited worlds would be as illogical as it would be morally repugnant. If we were to join a long-existent Galactic Club of intelligent life forms, we would surely be technologically junior to the vast majority of its established members and would not be in any position to "discover" or colonize them.

Nonetheless, the European Age of Exploration still remains a compelling era that warrants serious examination for what it can tell us about how a mix of curiosity, greed, science, and nationalism sent ships round the world and thereby changed the course of history. Yet it should not be looked at in total isolation from other episodes of maritime expansion. In particular, the Chinese expeditions into their South Seas and across the Indian Ocean that were underway at the very eve of Prince Henry's efforts demand parallel attention, especially for what they can tell us of the waxing and waning of governmental enthusiasm and support for costly projects of exploration (fig. 12.1). Although this short-lived chapter of Chinese maritime expansion predated by a few years the European one, let me begin this tale with the much better-documented and longer-lived efforts of successive European nations to explore the seas and gain wealth from distant lands.

In 1419 a Prince of Portugal was at loose ends. An expedition against the Moors of North Africa and then an assault upon their stronghold at Gibraltar had come to nought. Upon his return to Portugal, this younger brother of the King chose neither to settle in

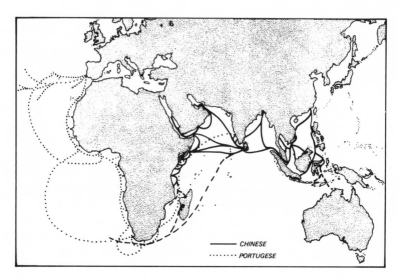

Fig. 12.1. Ocean voyages of the Chinese (solid = known or probable; dashed = conjectured); and the Portuguese (dotted) during the 15th century.

the royal court at Lisbon nor to take up residence on his lands in the north. Instead, he traveled to the very southwestern end of his country, to Sagres Point, where he rented quarters overlooking the windswept Atlantic. There, while brooding over his aborted crusade against the Moors, he began to formulate a plan that was to take the tiny caravels of Portugal down the west coast of Africa and eventually, after his death, around the tip of Africa and across the Indian Ocean to the rich ports of Asia. His immediate goal was to outflank the Moor, to sail down the African coast beyond the pale of Islam to gain direct access to gold and other African products and perhaps make contact with Christian princes rumored to live far to the south. To carry out this plan he sent expedition after expedition down the African coast and, to aid his captains to sail into unknown seas, he founded a school of navigation at Sagres. Thus did Prince Henry the Navigator open the European Age of Exploration—or at least so the story goes.

There is some exaggeration as well as legend here. Prince Henry was not a navigator. He never sailed on an exploring expedition. Nor did he found a school of navigation in the sense of a formal academy. Furthermore, he was not single-handedly responsible for all the Portuguese exploring initiatives of his day. Nonetheless, he was an inspirational leader in a process of systematic oceanic exploration, initiated by Portugal and followed in turn by a succession of European sea powers, which was to lead to global discovery and the erection of our current world system.

Actually, when Prince Henry returned from his crusading, Portuguese sailors had already begun to probe African waters on individual trading and fishing expeditions. What Prince Henry did was to transform this tentative and uncoordinated initiative into an organized effort. He encouraged sea captains to extend their explorations farther and farther to the south, and he himself financed scores of expeditions. When navigational problems of sailing down the African coast became an obstacle, Henry invited master astronomers and cartographers to Sagres to help develop means by which his captains could navigate farther into unknown waters. These pioneering efforts were continued after his death in 1460 by successive Portuguese monarchs who sponsored still more expeditions and fostered further technological developments needed to enable caravels to sail farther still into what the Portuguese poet Camões called "seas never before navigated."

How did the Portuguese do it? The ships they had at the beginning of the fifteenth century could easily sail down the African coast, running before the predominantly northerly winds. But the square sails of their sturdy little craft were not very useful for beating back home against those same northerly winds. The seeming timidity of sailors about venturing too far down the coast was based on a realistic assessment of the difficulties of returning as well as fear of the unknown. Mediterranean sailors had a craft that could sail well to windward, thanks to the lateen sail, a triangular sail introduced from the Indian Ocean into the Mediterranean by the Arabs. But these light, weatherly craft were ill adapted to rough Atlantic conditions. What the Portuguese seamen and shipwrights did was to combine Mediterranean and Atlantic features in one craft, to add to the robust Atlantic hull two or three masts rigged with lateen sails, and to steer the hybrid product with an Atlantic sternpost rudder. Thus was born the famous *caravela*, which carried the Portuguese farther down the African coast than they had ever sailed before, as they could now be assured of being able to sail home.

But the ideal way to sail home was not to tack back and forth keeping always within sight of land but to take one long tack hundreds of miles to the northwest until a fair wind was found that would allow a straight slant to Portugal. This brought on a navigational crisis. The compass, line, and dead reckoning might do for sailing in confined Mediterranean waters or for piloting along Europe's Atlantic coast, but to sail so far out of sight of land between two continents required some accurate way of estimating latitude. Otherwise seamen would

never really know how far down the African coast they had sailed or
when they had tacked far enough north to turn east for Portugal. This
is why Prince Henry invited those astronomers and cartographers to
Sagres—to adapt astronomical methods and instruments to the task of
determining latitude at sea and to draw up accurate charts of the
African coast and offshore islands.

Using the quadrant and later the astrolabe to measure the altitude of
Polaris gave the Portuguese seamen a direct indication of latitude, but
this method was useless during the day or when sailors, to their
horror, sailed too far south to see the North Star. In 1484 King João II
convened an international group of mathematical experts to work out
a practical method by which seamen could determine their latitude
from the sun. With the resultant "rule of the sun," captains were able
to probe still farther to the south, finally reaching a point past the
equator where the coast turned east—a long-anticipated sign that
Africa could be rounded and that the riches of Asia were within reach.

They were soon disappointed, for after several hundreds of miles
the coast turns south again, and what is more, the southeast trade
winds and the strong Benguela current flowing up the coast made it
extremely difficult for the caravels to make further progress, except
by tortuously tacking back and forth. The solution was to utilize the
trade wind regime of the Atlantic above and below the equator to sail
in a great arc far away from the coast and then south to pick up the
prevailing westerlies to sail before them, due east, past the tip of
Africa into the Indian Ocean. This is how in 1498 Vasco da Gama was
able to undertake that first direct trading voyage from Europe to India.

In the Indian Ocean the next technological innovation of the
Portuguese came into play. When da Gama displayed his wares before
the King of Calicut, the latter reputedly laughed, for the products of
Europe were of no interest to one who had the rich goods of Asia at
hand. He wanted gold and silver, but the Portuguese had only rough
cloth, scarlet hoods, metal washbasins, and other unimpressive
products of what was then the underdeveloped end of the Eurasian
land mass. But the Portuguese did have something unique in those
waters, which enabled them, in just a few years, to establish a line of
trading forts extending throughout Asia. On the decks of their stoutly
built ships were mounted cannon, which the aggressive Portuguese
used to bombard ports and blast poorly armed rival fleets out of the
water and thus bully their way into the rich trade of the Orient.

The Portuguese had sailed into a power vacuum, but one only
recently created. Had they reached India when Prince Henry first
began to contemplate the sea at Sagres Point, they would have met

formidable rivals—master seamen from the biggest nation in the world, sailing in the largest ships and fleets yet seen on the ocean.

China had long been a scene of maritime innovation. The compass and the sternpost rudder, both essential to the development of European long-range voyaging, were first developed by the Chinese. As early as the third century A.D. they were sailing large, multimasted junks driven by sophisticated mat and bamboo lug sails which enabled them to point well to windward. What is more, the ships were built on an entirely different principle from Western craft. Instead of being made up of a keel and ribs clad in planking, the Chinese junks were made of watertight compartments, which gave them a structural integrity and, as attested by those medieval travelers Marco Polo and Ibn Battutah, a degree of comfort far surpassing Western vessels of the time.

Chinese maritime expansion was at its peak when the Portuguese were first timidly venturing down the African coast. Between 1405 and 1433 the Ming Dynasty emperors sent out seven "Treasure Fleets" to the islands of southeast Asia, to India and Sri Lanka, and to Arabia and Africa. One of these great fleets was reportedly composed of some 62 major ships that carried more than 27,000 officers and men. The junks of this era could be truly huge, larger ones averaging 450 feet in length, with a beam of 100 feet and room for a crew of some 500 men. Not just one but a whole collection of Prince Henry's little caravels could have been laid athwart the deck of one of these huge junks with room to spare.

The nature of these Chinese expeditions into the Indian Ocean during the Treasure Fleet era was markedly different from that of the Portuguese expeditions that followed. Whereas the latter were aggressively commercial and carried out in a crusading piratical style, the Chinese ventures were, as the historian Joseph Needham puts it, "the well-disciplined naval operations of an enormous feudal-bureaucratic State." Although some traders went along on the junks, and more probably followed in their wake, the primary impetus was governmental and diplomatic. The fleets were showing the flag around the Indian Ocean, bestowing gifts upon local kings and rulers and, in return, demanding tribute and submission to their Emperor. Although the Chinese brooked no opposition, only three serious conflicts on all the voyages marred these otherwise remarkably peaceful ventures in international and intercultural relations. The Chinese certainly did not prey, as the Portuguese later did, on Arab traders and trading states. They had no quarrel with Islam or any other religion. After all, their Admiral, the famous Cheng Ho, was himself a Muslim whose father had made the pilgrimage to Mecca.

The expeditions also had a natural history component: the collection of gems, minerals, drugs, and exotic plants and animals, including the giraffe from Africa, which was highly prized because it resembled a mythical beast whose appearance on Earth was supposed to signal that a just Emperor was on the throne.

Yet when Vasco da Gama rounded the Cape of Good Hope in 1497, he encountered no Chinese Treasure Fleets, no huge junks to awe or oppose him. By then the Chinese had withdrawn from the Indian Ocean, in fact practically from the sea itself. The junks from the last Treasure Fleet were broken up or allowed to rot. No new ships were built and no more fleets were dispatched to far-off lands. China had abruptly turned inward to the point of forbidding private overseas trading ventures. By 1500 it was even a capital offense to build a seagoing junk with more than two masts.

What happened? Why did China renounce the sea? Why was it left to the tiny caravels of Portugal to open the era of global exploration and to establish the first world sea-borne empire? It was not a matter of population or gross national product, for Portugal was but a small, underdeveloped country of less than a million inhabitants compared with China, then the world's largest and richest nation. Nor was technology in itself the crucial variable. When the Portuguese were just starting to develop an overseas maritime technology, one based largely on features that had evolved elsewhere, including the compass and sternpost rudder which were of ultimate Chinese derivation, the Chinese were already sailing over two oceans in sophisticated, compass-navigated junks, which were the fruit of a thousand or more years of Chinese maritime development. Had the Chinese kept going, had they rounded the Cape of Good Hope, sailed north, then anchored off Lisbon, explored the Thames, or cruised the Mediterranean, all of which they were technically capable of doing, how different the modern world might have turned out. But the Chinese withdrew just as, unbeknown to them, the Portuguese began their advance.

The villains of this Chinese puzzle, the ones who stopped Chinese maritime expansion, were the official bureaucrats of the State, the Confucian Mandarins. The tragic heroes, at least to those who might wish the Chinese initiative had been allowed to run its course, were the Ming Emperor Yung-Lo and his Admiral Cheng Ho, a most unusual champion of overseas exploration. The Admiral was a eunuch, born in Yunnan, the last Mongol stronghold in China, captured as a boy by the victorious forces of the son of newly emergent Ming Emperor, then castrated and placed into the service of the conquering Prince. It was the latter, when he became the Yung-Lo Emperor, who launched the first Treasure Fleets. In so doing he turned for leadership not to the official bureaucracy but to his personal eunuchs and in particular to Cheng Ho who directed the whole enterprise from start to finish (and probably had a major hand in its conception) and who personally commanded many of the voyages.

This did not sit well with the Mandarin civil servants. However, they had to bide their time until Yung-Lo died in 1424. Then one of their number, Hsia Yuan-Chi, emerged as the champion of retrenchment against this waste of money in useless exploration. In urging that the fleet movements be stopped immediately, he wrote, in words that come down to us, "If there are any ships already anchored at Fukien or Thai-tshang they must at once return to Nanking, and all building of seagoing ships for intercourse with barbarian countries is to cease forthwith." But before the year was out, the new Emperor died and his successor ignored this bureaucratic urging and launched the seventh Treasure Fleet. Alas, this was the final sailing, for this Emperor died soon thereafter, as did Cheng Ho, and the Imperial policy was reversed once more, this time permanently.

It is easy to see this as but one skirmish in the long-standing battle between the Confucian bureaucrats and the Imperial eunuchs. But bureaucratic opposition to overseas ventures went much deeper than this rivalry. As the classically educated sons of the country landlord class, the Confucian bureaucrats had always tended to look askance at any intercourse with foreign countries. They had no interest in other countries, nor in their luxuries which, according to the Confucian ethic of scholarly austerity, were wrong in themselves. Besides, they could argue that since China could produce all the food, clothing, and craft products she needed, there was no need for overseas trade. Above all, as Joseph Needham explains, "the Grand Fleet of the Treasure-ships swallowed up funds which, in the view of all right-thinking bureaucrats, would be much better spent on water-conservancy projects for the farmers' needs, or in agrarian financing, 'ever-normal granaries' and the like."

This abrupt termination of overseas voyaging was not, of course, an isolated incident going against the overall trend of Chinese history. It occurred during a time when a highly conventionalized version of neo-Confucianism was developing, one that discouraged interest in science and technology in favor of sterile introspection. But to inquire further into why China turned inward at this time would lead us into the thorny and often-discussed questions of why China did not have its own Renaissance and scientific revolution, its own transition to capitalism, as well as its own era of global colonial expansion—in other words, why China did not develop as the West. It is more useful for us to examine more closely just why the Europeans expanded over the seas.

Were the early European voyages of exploration primarily attempts to investigate unknown portions of the world, or was the motivation basically economic? Modern historians emphasize economic causes, and their analyses go far beyond the usual recounting of how Portugal, Spain, and other nations sought sea routes to obtain the riches of Asia.

According to a current, neo-Marxist view, the Age of Exploration was a response to crises conditions in Europe. By the fifteenth century feudalism was at the point of collapse. All the surplus possible was already being squeezed from the peasantry, the technological limits to the expansion of agricultural production had been reached, and the worsening climate of the coming Little Ice Age was making a bad situation worse. Europe had to expand out of its geographical bounds, had to obtain new sources of food, raw materials, and gold and silver for trade or else face economic ruin and anarchy. The voyages of exploration constituted the reconnaissance necessary to establish direct trade routes to distant sources of supply and, as it turned out, to open up new lands for direct European colonization. The Age of Exploration saved Europe by leading directly to the erection of the world capitalist system dominated by the West.

But this is simply too much economic determinism. It leaves out the crucial motivational factor exemplified in the drive of visionary explorers to prove their theories, and above all it ignores a critical development then occurring in the West. Without the contributions from the developing scientific and technological movement of the day, Europe could never have resolved its crises through overseas expansion. Exploration and science were intimately linked from the beginning. In fact, the recruitment by Portugal of mathematicians, astronomers, and other scientists to help seamen navigate uncharted waters was one of the first, if not the first, major instance in which science was applied systematically to solve a practical problem. Clearly, the technological attitude then developing in the West, with its extreme readiness to apply scientific findings to anything from navigation to gunnery, combined with the search for new sources of wealth to produce European overseas expansion.

This potent combination of a new technological-scientific approach and nascent capitalism propelled one European nation after another into the race to explore the world and exploit the new routes and lands discovered. Portugal led the pack, largely because of three factors: its location, jutting out into the Atlantic where favorable wind and currents allowed easy access to West Africa, its high degree of national unity, then unique in Europe, which meant that exploration could more easily be made a governmental concern, and its Iberian heritage of alternately fighting and trading with the Moors, which promoted an extra-European outlook. Neighboring Spain, the next nation to focus on overseas exploration, largely shared Portugal's advantages, although it lagged in national unity. However, neither of these two leaders in Europe's expansion was a center for science and capitalism. No matter. They could hire the scientists needed to develop navigational techniques, and they quickly became a magnet for visionary seamen from around the Mediterranean who sought backing for their voyages, including that most famous son of Genoa, Christopher Columbus. In addition, Europe's bankers, particularly the Italians and South Germans, were attracted by the developments on the peninsula and they offered their capital for even the most speculative of ventures. For example, the German House of Fuggar, through its Spanish subsidiary, invested heavily in Magellan's voyage and supposedly even made a small profit from the spices brought back on the one ship out of five that completed that first circumnavigation. Later the bankers of Genoa, blocked from the rich eastern Mediterranean trade by Venice, bankrolled Spain's exploitation of the New World.

But Iberian domination was not to last. Spain had hardly established its New World empire when it started its slow decline. Portugal's fall from world economic power was much more rapid. "Greater robbers came to prey upon the lesser," is how Joseph Needham characterizes Portugal's demise as the European trading power in Asia. Among the awakening European maritime powers of the sixteenth and seventeenth centuries, it was the Dutch who emerged as the greatest of the robbers. Upon casting off Spanish rule, this newly independent nation sent its ships into Asian waters where they out-gunned and out-traded the Portuguese, driving them from the Spice Islands, Sri Lankan ports, and other trading centers. The Dutch now had the best ships, seamen, navigators, and gunners and employed to great effect an economic innovation: the joint stock company. Previously, trading expeditions had typically been financed one at a time by individuals or ad hoc partnerships, with the capital and profits often divided at voyage end. The newly developed joint stock company could give the continuity needed to regularize and expand trade. The English saw this early and organized East India Company in 1600. But the Dutch gained the commercial edge when two years later their government set up the Dutch East India Company, a government-sponsored monopoly that amalgamated all their private traders into a vast and effective corporate structure, with military and governmental functions, which was to make Holland the global trading power of the seventeenth century and Amsterdam the commercial capital of the world.

In the following two centuries it was England's turn. With her superb navy she vanquished Spanish, French, and Dutch foes and, among other things, inaugurated the era of truly scientific exploration with the magnificent voyages of Captain James Cook. With her commercial acumen, growing industrial capability, and colonizing genius, England was able to fashion a world empire that eclipsed all previous attempts, but which had a limited life span like all the rest.

It is tempting to try to apply these insights from ocean exploration and colonization to the coming expansion into space. The contrast between Portugal and China certainly suggests that motivation, not mere size or wealth, is the crucial ingredient for undertaking exploration beyond the known world. In addition, the rise and fall of Portugal, Spain, Holland, and Britain suggests that no one nation—no matter how wealthy or technologically advanced it might seem at the time—will be able to maintain the political commitment, entrepreneurial drive, and technological innovativeness to stay perpetually ahead in space.

The eroding lead of the United States in space provides a case in point. Is the U.S. space program in danger of going the way of the Ming Dynasty navy? Are the unused Apollo rockets from canceled moon missions harbingers—like the rotting hulks of the great junks of the abandoned Treasure Fleets—of a fall from greatness? Had the United States totally abandoned space after the triumphant moon landings, the analogy would be utterly compelling. However, despite budget cuts, a vigorous program of unmanned planetary exploration followed in the 1970s, as did the development of the space shuttle. Still, without a formal commitment to settle space, we must wonder how prominent Americans will be in fulfilling the visions of Tsiolkowsky, von Braun, O'Neill, and other modern-day analogs of Cheng Ho and Prince Henry.

In fact, if we were to carry the maritime analogy to its extreme, there would be a temptation to dismiss both the United States and Russia and nominate Japan to be the preeminent space power for the next century. After all, once upon a time another small island nation with a tremendous capacity for industrial innovativeness, production, and export grabbed the world maritime lead away from her larger continental neighbors.

However, such thinking is probably too time- and Earthbound. I suggest that for the future in space we have to look beyond competition between terrestrial states and the truism that nations rise and fall. (Here my optimism shows, for I think superpower competition will be attenuated in the vastness of space.) The real lesson may be that just as global oceanic expansion called for new forms of economic and political organization, so will space settlement require another round of social as well as technological innovation. Although the search for wealth by sea captains, bankers, and princes helped drive maritime expansion, we can hardly say that the current world capitalist system caused it. In an important sense, our global economic system is a necessary outcome of that expansion. The Age of Exploration produced a new way of ordering economic relations between peoples and nations previously separated by the 70 percent of Earth's surface that is water. Should not the infinitely greater challenge of settling space and of tapping the resources of the entire solar system (and of star systems beyond) call for radically new social structures?

Just as modern capitalism, and the political arrangements that support it, evolved out of feudalism, so in space some new form of organization will evolve out of the current world capitalist order. I do not mean socialism, communism, or any other reform movement of capitalism. Rather, I think that some entirely new way of ordering economic and political life must develop when—barring nuclear holocaust or some other life-destroying disaster—our descendants settle in the vastness of space and exploit the almost unimaginable resources available there.

References

Cortesao, Armando. 1966. Note sur les Origins de la Navigation Astronomique au Portugal. In M. Mollat and P. Adam, eds., *Les Aspects Internationaux de la Découverte Océanique aux XVe et XVIe Siècles*. S.E.V.P.E.N., Paris.

Duyvendak, J. J. L. 1949. *China's Discovery of Africa*. Arthur Probstein, London.

Godinho, Vitorino Magalhaes. 1965. *Os Descobrimentos e a Economia Mundial*. 2 vols. Arcadia, Lisbon.

Levenson, Joseph R. 1967. *European Expansion and the Counter-Expansion of Asia, 1300-1600*. Prentice-Hall, Englewood Cliffs, N.J.

Lo, Jung-pang. 1958. The Decline of the Early Ming Navy. *Oriens Extremus* 5: 149-168.

Morison, Samuel Eliot. 1974. *The European Discovery of America: The Southern Voyages A.D. 1492-1616*. Oxford University Press, New York.

Needham, Joseph. 1971. *Science and Civilization in China*. Vol. 4 *Physics and Physical Technology*. Pt. 3, Civil engineering and nautics. Cambridge University Press, Cambridge.

Parry, John Horace. 1964. *The Age of Reconnaisance*. New American Library, New York.

Ure, John. 1977. *Prince Henry the Navigator*. Constable, London.

Wallerstein, Immanuel. 1974. *The Modern World-System*. Academic Press, New York.

"When I was a boy, there was but one permanent ambition among my comrades in our village on the west bank of the Mississippi River. That was, to be a steamboatman."

—*Mark Twain*

"History repeats itself. That's the one thing wrong with history."

—*Clarence Darrow*

✳13

Alfred W. Crosby

LIFE (WITH ALL ITS PROBLEMS) IN SPACE

The most dangerous enemy of humans beyond Earth's atmosphere and beyond the moon's orbit in the long term will be the implacably hostile environment. That will be a problem for the engineers, and their record thus far suggests that although there will be a few awful accidents, there will be *very* few. The next most dangerous enemy in space in the immediately foreseeable future will be humans themselves and the irrepressible organisms they will carry with them.

Engineers are successful because they apply their skills to quantitatively predictable phenomena. Ecologists, zoologists, botanists, and epidemiologists are not so successful, because they deal with life, which grows, changes, declines, plunges to extinction, scrambles for survival, and rises and plays the phoenix again and again on a merry-go-round that spins in all four dimensions and on a varying number of axles all at the same time. Space engineers are well prepared for their tasks, but not so the space ecologists. In fact, are there any scientists specializing in what will happen to life forms in the hermetic hermitages of space?

Let us begin with kindergarten ecology. An ecosystem is an active collection of interdependent organisms. They depend on one another for food and shelter and much more. Ultimately, all organisms within a given system are in symbiotic relationship. They depend largely on one another even for death, which is so necessary to the continued health of the system.

Species, such as humans, crabgrass, and tobacco mosaic viruses reproduce many times more than are needed to simply replace the parents numerically. Every ecosystem is in danger of being smothered by the reproductive excesses of its members, but is usually saved from this fate by the appetites of its members, that is, the participating organisms obligingly eat up each other's excessive offspring.

Each ecosystem (and, by extension, each world) is a matter of shifting balances between its participants. Each ecosystem is rather like a tightrope walker: his pole wags up and down, but the walker is stable. He is in a state of what biologists call homeostasis, or what the poet, addicted to conceits, might call dynamic stasis. He continually readjusts his balance to stay balanced.

What happens when two tightrope walkers touch or ever so slightly interfere with the serene bobbing of each other's poles? They fall or regain balance only by violent exertion. What will happen when bands of humans and associated life forms, separated in space long enough to establish independent states of homeostasis—perhaps even long enough to start down separate paths of speciation—contact one another?

We are getting too abstract. Let us approach this problem from another angle. An individual organism does not exist in space and time in general, but in a specific place at a specific time. It does not exist *in the long run*; it exists *in the short run*. Any period longer than that is simply a series of short runs. Within a specific short run, it works out a specific *modus vivendi* in a specific environment with specific other organisms, macro and micro. Shift this protagonist—human, bacterium, or what-have-you—from one short run to another and you may be shifting from a game of chess to a rugby scrum. In fact, you may bring about that appalling transformation simply by the act of arriving unannounced in somebody else's chess tournament.

An example may help. Henry VIII reigned in England in the first half of the sixteenth century, and he worried about Francis I of France and not about kings in general. He married Katherine-of-Aragon-through-Catherine-Parr, and not (not even Henry) women in general. He ate wheat and beef, not food in general. He was subject to specific variants of specific diseases, plague in particular, and not to disease in general. For instance, his immune system was as unacquainted with—that is, was no more fit to defend him against—yellow fever than he was to conduct guided tours of the Niger delta.

Henry VIII was a thorough Renaissance Englishman, as hermetically sealed in his time and space as anyone could be in a spaceship or colony. If he had access to a time machine and if he had used one to join us here and now, he would find our clothes silly, our English nearly unintelligible, our food weird, our cattle and sheep all giants, the variety of our breeds of dogs nearly infinite, and our women disobedient. He would fall sick with a number of our diseases because his immune system would be no better prepared to protect him from our pathogens than he himself, an excellent musician, would be to play music on instruments tuned according to the system of equal temperament celebrated by Johann Sebastian Bach in the eighteenth-century preludes and fugues of his "Well-Tempered Clavier."

The bacterial flora of Henry's alimentary canal would be different from ours, and as ours invaded his ample gut, he would become the first time traveler to suffer from the *turistas*. He would simultaneously or shortly after fall victim to invasions by such constantly altering viruses as those of the common cold and influenza. He would be liable to infection by the bacteria of Legionnaires' disease, which some experts suggest came to its present state of virulence in the special environment of our air conditioners, which, of course, did not exist in the sixteenth century. Herpes and AIDS would, as you might expect, find him defenseless. King Henry, a Renaissance organism (or from the point of view of microlife, a Renaissance ecosystem), would be out of place—or in this case out of time—and susceptible to all kinds of alien infections and infestations.

In addition, he would be a threat to us. He could be carrying fleas—the sixteenth century was not fastidious about cleanliness—and the fleas could be carrying the bacteria of plague, one of the chief killers of his age. We have antibiotics to cure plague, but how could we be sure they would be effective against strains of microlife 450 years out of date? Henry might be carrying the spirochetes of syphilis ("the frensshe pockes"), a disease slow to wreak havoc in the individual human body in our time but a kind of galloping leprosy in the sixteenth century, as you might expect of a spanking-new malady that had first been noted in Europe in the 1490s. Henry might bring to our century the organisms that caused the English Sweats, an infection that killed many in his lifetime, organisms for which we probably have no defenses in place because the disease has not appeared—not at least in its virulent form—since then.

But this is all fiction. We have resorted to a time machine to assure that our traveler moves from one short-run situation to what would clearly be a very different one. Is there no history, no example from reality, to illustrate the kinds of danger that space dwellers might present to one another? There are: History is full of them. We have, you see, done all this launching out into space before.

Humanity taught itself to live in environments for which it was not and still is not genetically adapted—first by learning how to make tools in Africa and then, after straying from the African Eden to the cooler climes of Europe and Asia, how to make clothing and use fire. However, although this diffusion into new environments spread humanity far and wide, the great majority of human groups were in at least occasional contact with other groups, even though the intervals between contacts might be a matter of generations.

After a long while *Homo sapiens*, having sifted through Africa and Eurasia, went on elsewhere. During the later millennia of the Pleistocene, when much of the world's water was fixed in continental glaciers and the levels of the oceans were lower than they are today, some humans gradually, over a period of generations, migrated from Siberia into Alaska and thence all the way to Tierra del Fuego. Others moved southeast from Asia, skipping over the narrow gaps between the islands of Indonesia, arrived in New Guinea, and from there walked into Australia. By the end of the Pleistocene, humans were by all measures the most widely distributed of all large animals. They had occupied, at least thinly, all the habitable parts of the globe except Madagascar, New Zealand, and smaller islands.

Then the world heated up, the glaciers retreated, the oceans rose, isolating the avant-garde of the human race in the Americas and Australia. A few millennia later the Polynesians, a sort of Johnny-come-lately avant-garde, attained similar isolation by sailing out and settling the atolls and volcanic islands of the pelagic Pacific.

Their isolation and that of the first Americans and Australians would have lasted until oceans fell again or tectonic drift brought continents thumping up against each other—but for human inventiveness. Our teeth and claws are not impressive, but our brains make up for their inadequacies. As of, say, A.D. 1000 there was little if any contact across the Atlantic or Pacific. The various human cultures on opposite sides of those oceans had grown up in isolation, as had the flora and fauna, macro and micro, and the various systems of biological homeostasis.

During the period Europeans call the Middle Ages, peoples of the Old World made important advances in transportation technology. Stirrups, the horse collar, and such may seem minor advances from our perspective, but they were of enormous importance in an age

when power was chiefly muscular. Naval transportation technology
advanced too. Hull design, sail and rigging design, and navigational
instruments and techniques improved more in a few centuries than in
the previous two millennia. The results are perhaps the most awesome
example of the influence of technology of all time.

The three great naval traditions of the Old World were those of
Europe, the Middle East, including the Muslim areas of India, and
China. The Muslims were great sailors, coursing back and forth,
century after century, between the Middle East and India at the center
and East Africa and Southeast Asia at the periphery. But they rarely
sailed outside the Indian Ocean and East Asian seas and did not go on
to conquer the Atlantic and Pacific. Why? Possibly because the
monsoons, sweeping out of central Asia in winter and back in from the
oceans in summer, solved most of their sailing and navigational
problems for free, so to speak. After 1000 or 1100 or so, the Chinese
and, lagging somewhat behind, the Europeans took over unequivocal
leadership in the arts and sciences of nautics.

Chinese navigation right up to the sixteenth century was at least as
good as European, and their vessels larger and probably more seaworthy. They had sternpost rudders instead of inefficient steering
oars; their hulls slipped through the water with less turbulence and
drag and were squarer in cross section, providing a more stable carrier
for more freight than the Occidental equivalent. The Chinese divided
their hulls into watertight compartments, giving their vessels the
multiple lives Westerners claim for their cats. More important, their
bamboo-stiffened sails, capable of turning a nearly rigid edge into the
wind, were better than any Western equivalent for beating into the
wind.

The Europeans were behind, but making advances too. They
adopted the lateen sail from the Arabs, which enabled them also to
tack into the wind; and the sternpost rudder, possibly taken indirectly
from Chinese examples; and schemes of multiple masts and sails, also
possibly of Chinese provenance. The compass arrived, probably from
the Far East, by the thirteenth century.

The Age of Exploration, of transoceanic voyaging, was about to
begin. The first and greatest admiral of the age, in terms of the size of
his fleets, was Cheng Ho, chief eunuch to the emperor of China. The
eunuch built fleets of scores of ships, many of the vessels among the
largest in the world, fleets capable of carrying tens of thousands of
officers and men. He personally led or organized and dispatched
seven expeditions between 1405 and 1433. They sailed to Siam, the
East Indies, and beyond to India, Ormuz, Aden, and East Africa—to
Malindi where a single lifetime later Vasco da Gama would pick up a

pilot to guide him across to India. Cheng Ho's last expedition returned to China in 1433, and two years later his emperor died. The conservative Confucianists and rural landlords who dominated the courts of succeeding emperors directed China's attention back to her domestic and dry land affairs. China—on the brink of *everything*—did an about-face. Europe took over the lead on the high seas.

In 1291 the Vivaldi brothers from Genoa sailed out of the Mediterranean with a plan to circumnavigate Africa and were never heard of again. European naval technologies were not yet up to such a task, but the improvements came and with them the kind of successes of which the Vivaldis had dreamed. Italian, Mallorcan, Breton, and Iberian sailors, navigators, and shipbuilders learned the rudiments of oceanic seamanship in the fourteenth and fifteenth centuries and discovered or rediscovered the Canaries, Madeira, and the Azores. In 1488 Dias found the way around Africa, and in 1492 Columbus the way to America. In 1521 the survivors of Magellan's fleet completed the very first circumnavigation of the globe. The continents, which had been connected by land in the Pleistocene, were reunited with a sailmaker's needle.

The ecological facet of this reunification was by far its most spectacular. The first examples of the wild oscillations that commonly follow upon contact between previously isolated ecosystems actually predate the voyages of Columbus. Early in the fifteenth century his future father-in-law innocently brought the first rabbits to Porto Santo, the companion island of Madeira, and in a year or so the colonists had to get out because multitudes of rabbits were eating everything they planted. In the Canary Islands the indigenes, called Guanches, who numbered in the tens of thousands in 1400, were plunging toward extinction by 1492. The cause of their demise was not so much European brutality, of which there was plenty, but alien diseases. *La peste* and *modorra*, whatever they may be precisely, swept through the Guanche populations in the 1480s and 1490s.

These happenings were omens of what was to follow. The European arrival in the Americas set off what are the most awesome ecological events in recorded history. Diseases unknown in the New World—like smallpox, measles, yellow fever, and malaria—rolled through the two continents, and reinforced by the conquerors' penchant for intentional and unintentional genocide, they produced the greatest demographic catastrophe in the human record. Demographic historians in general agree that American Indian populations usually declined 50 to 100 percent in the first century after full contact with Europeans and Africans. In central Mexico, where we have the best statistics, this is clearly the case. Smallpox was the chief killer, at least initially, and its impact helps to explain just how and why the conquistadores were so successful. Smallpox accompanied Cortes into the Aztec Empire and preceded Pizarro into the Incan Empire, killing large proportions—no one knows how large—of the peoples of these societies. Death rates of a third and a half are often cited and are probably not far from the truth. Other new diseases undoubtedly operated alongside smallpox, as did starvation and war. Similar "virgin soil epidemics" in Polynesia during the last century and in Alaska and the hinterlands of Venezuela and Brazil in our century have produced mortality rates almost as impressive.

While such epidemics transfigured the history of humanity in the Americas, oscillations similar in magnitude were sweeping other species. In 1492 no horses, cattle, sheep, or pigs lived in the New World. Within a hundred years there were millions. Northern Mexico and the pampas of Argentina and Uruguay ran with Old World quadrupeds by the tens of millions, animals just as wild as their ancestors had been a few thousand years before. The story goes on in our time. In the 1890s a few score European starlings were released in New York's Central Park; today they range from coast to coast and are estimated at 100 million in number.

Similar events mark the histories of Australia and New Zealand. Smallpox slaughtered Australian aborigines and tuberculosis the New Zealand Maori. In the mid-nineteenth century the first European rabbits were introduced into Australia, and in a hundred years they numbered a half billion. In New Zealand dozens of Old World species (and Australian as well) have become acclimated and have overwhelmed the native creatures over wide areas.

Such biological spectaculars have not been restricted to ecosystems separated by oceans. All that is required is a high degree of isolation and then an abrupt opening of contact. For an example, let us turn to Africa in the last years of the nineteenth century. European imperialists had conquered the Americas and Australia with relative ease, but sub-Saharan Africa, populated by peoples with iron weapons and a strong military tradition and defended by yellow fever, malaria, amoebic dysentery, and such, remained independent all the way to 1850 and beyond. Then advances in sanitation and medical science enabled Europeans to penetrate tropical Africa with reasonable safety. Historians call what followed "carving the African melon."

Europe entered Africa, and the ecological disruptions were severe. The most spectacular of these pertain to rinderpest, a commonly fatal viral disease of hooved animals, especially cattle, which apparently was entirely absent from Africa south of the Sahara. It arrived in the 1880s, probably with the cattle brought from Russian and other Black

Sea ports to feed the British Army in Sudan and/or with cattle the Italians imported from India and Aden to provide for their invasion of Ethiopia. The disease was first reported in Somaliland in 1889, reached Uganda the next year, Lake Nyasa in 1892, and raced on south and west. "Never before in the memory of man," wrote one witness, "or by the voice of tradition had cattle died in such numbers." Peoples dependent on them, like the Masai, starved in large numbers. Wild game, the usual source of food during famine, was not there to save the Masai this time because the African ungulates fell before the new disease in windrows. Two-thirds of the Masai died, says Oscar Baumann, who was there to see.

Bringing previously isolated ecosystems together is much like flicking a cigarette lighter near open containers of gasoline. Some of the time nothing will happen. Some of the time the fumes will ignite and blow your head off. Ecological explosions will not always happen if societies and ecosystems which we plant beyond Earth are carelessly brought together after long intervals of separation, but such detonations will take place some times, and they will be very difficult to deal with. Measures taken to control the propagation of one or several kinds of organisms will be dangerous to all—especially in situations in which airing out the room will not be a corrective action with much appeal to the rational traveler.

Interstellar migration will maximize the rates of differentiation between the groups of migrants: humans, plants, animals, and microlife. Such groups will develop from small initial stocks, maximizing the founder effect, that is, the inevitable initial differences will magnify over the generations. These stocks will propagate in quite possibly very different environments, and will very probably evolve in hermetic isolation from one another. The moments when they meet will not only be times of extreme opportunity but also of extreme danger for all organisms involved. Humans, who reproduce much more slowly and less often than the other organisms they will intentionally and unintentionally take with them—humans whose immune systems react slowly to assaults by unfamiliar enemies—will be hard put to survive the biological chaos that will often follow.

We are sometimes told by optimists that such disasters can be avoided. One, space travelers will all be inoculated against all the standard diseases, and vaccines will be taken along into the void. But effective vaccines do not exist for every disease—herpes, for instance—and pathogens evolve; the flu vaccine of 1963 gives little protection against the flu virus of 1983. Furthermore, we are now learning that there are infections that take decades to manifest themselves. Providing defenses against them requires miracles of foresight.

It may be true that viral DNA can lurk unmanifested within the nuclei of reproductive cells for entire lifetimes, be passed on to descendants, and then declare independence and cause disease and dysfunction. The whole mystery of the relationship between "our" genetic materials and those bits of genetic information we call viruses may contain jolly surprises for humans locked in reluctant embrace and sealed in tiny quarters whizzing through interstellar vacuum.

The optimists assure us that extreme care will be taken to restrict space travelers to humans plus just those species that humans want as sources of food, oxygen, and companionship and to recycle waste. But such measures will not preclude microstowaways because there is no way to fully sterilize living things inside and out without killing them. Large organisms, like humans, pigs, and cabbages, are gardens of microorganisms, some of which are indeed very likely essential to their health. In addition, we have to consider that a large organism, absolutely free of bacterial occupants, would be an open invitation to new occupants. Such an organism *would be occupied by the first microlife to come along*, and the odds are in favor of the occupation being rated by the occupied as a severe infection.

Species, said Charles Darwin in the nineteenth century, are not stable, and they certainly will not be in the twenty-first century. We are assured that space travel will provide plenty of time for genetic change, during which, incidentally, the radiation levels are likely to be higher than on Earth, accelerating that change. Preparing for mutations—what will that be like? Surely, nourishing, fast growing, tasty molds, likely candidates for space travel, can under certain conditions produce offspring that will take happily to growing on and in the organic plastics that compose or enclose much of our electronic and other gear.

Humans over time can take measures to bring such threats under control but when space societies meet, such threats will sometimes rise up meteorically. Your staphylococci is for you a cause of pimples; for me and my comrades a cause of confluent rashes, massive loss of fluids, and dangerous secondary infections.

We can prepare for the predictable threats of the void—the vacuum, the cold, the heat, the radiation. We will find it very difficult to prepare for the threats that arise from the organic nature of ourselves and of the forms of life we will take with us. Put as simply as possible, the problem is this: We can shake loose from our home planet, but we most assuredly cannot leave DNA behind us.

References

Ciba Foundation. 1976. *Health and Disease in Tribal Societies.* Symposium 49, new series. Amsterdam, Elsevier/Excerpta Medica/North-Holland.

Crosby, Alfred W. 1972. *The Columbian exchange: Biological and cultural consequences of 1492.* Greenwood Press, Westport, Conn.

Crosby, Alfred W. 1978. Ecological Imperialism: The Overseas Migration of Western Europeans as a Biological Phenomenon. *Texas Quarterly* 21 (Spring), 10-22.

Kjekshus, Helge. 1977. *Ecology control and economic development in East African History: the Case of Tanganyika, 1850-1950.* William Heinemann, London.

McNeill, William H. 1976. *Plagues and peoples.* Anchor Press/Doubleday, Garden City, N.Y.

Wolfe, Robert J. 1982. Alaska's Great Sickness, 1900: An Epidemic of Measles and Influenza in a Virgin Soil Population. *Proceedings of the American Philosophical Society 126* (April 8): 91-121.

✻ 14

Nancy Makepeace Tanner

INTERSTELLAR MIGRATIONS:
THE BEGINNINGS OF FAMILIAR PROCESS
IN A NEW CONTEXT

Major migrations have occurred at various points in human evolution, and more recent phases of human history also exhibit a variety of ways in which outward movement, expansion, and migration have occurred. Can these past migratory experiences give us any clues for use in the future as we begin to move off the globe and, eventually, beyond the Solar System? In this essay I will try to address that question by referring to the anthropological record, drawing in particular on my own research on human evolution and intercultural relations.

Anthropology is a synthesis: It looks at past human evolution through fossils and ancient tools, as well as through comparison with our primate cousins; and it looks at the variety of currently existing human innovations—technological, social, and cultural—by comparing a wide variety of human cultures. My own research has focused on two quite different aspects of anthropology. I have worked extensively on early human evolution with regard to both our ape ancestors and the early hominids. Specifically, I have investigated the chimpanzee as a model for reconstructing features of our nearest ape ancestor and have focused also on the transition from ape to hominid and the role food gathering played in this development. In addition, I have spent four years, in four separate field trips, as a social anthropologist doing cross-cultural research among the Minangkabau people of West Sumatra, Indonesia—a matrilineal, Islamic group that is both traditional and very modern, depending upon what aspects are being considered. The Minangkabau are particularly intriguing for colonization theorists, for they are a people of great cultural pride who have traveled widely over much of the world, yet also a people whose behavior toward other groups they encounter is virtually the antithesis of that usually associated with colonialism. By examining the evolution and spread of early hominids and by comparing Minangkabau and European intercultural styles, I hope to show how insights from the anthropological record can be related to future issues in space migration.

Migratory Points in Human Evolution

The Transition from Ape to Human: A Movement from Jungle to Savanna

Humans are very closely related to African apes. During the past decade evidence has continued to grow regarding the period of time from which the human and ape lines diverged. With more evidence the divergence time gets shorter—for African apes in general and chimpanzees in particular. The fossil record is not yet conclusive: As yet there are very few fossils for either apes or hominids between 8 and 4 million years ago (Mya). However comparison of protein molecules, DNA, and chromosomes among humans, apes, and monkeys helps to fill in the gap. Amino acid comparisons place the separation of the hominid line from African apes (chimpanzees and gorillas) later than from Asian apes (orangutan and gibbon). Even more fascinating is chromosomal evidence, which, according to J. J. Yunis and O. Prakash, suggests "the existence of a progenitor of chimpanzee and man *after* the divergence of gorilla." Particularly interesting are Sibley and Ahlquist's extensive recent ape-human comparisons by DNA-DNA hybridization. They too find chimpanzees to be the ape most closely related to humans.

In other words, chimpanzees are very close kin. They provide us with the best available model—whether from a genetic, molecular, anatomical, or behavioral perspective—for our ape ancestor. The similarities between common and pygmy chimpanzee species indicate chimpanzees have remained fairly conservative, perhaps retaining a high degree of similarity to the African ape ancestor of humans and gorillas as well as to themselves. In comparison, humans have changed considerably: Bipedalism and much bigger brains characterize our anatomy. Also, culture, including speech, is central to the human adaptation. Yet at the very beginning of human evolution, our ancestors were apes, adventurous apes to be sure, but probably not very different from the chimpanzee.

The first major movement in our evolutionary history was probably made by a chimpanzeelike ape ancestor moving out of the African tropical rain forest into the East African savanna—a mosaic environment of grasslands with scattered patches of forest. Initially, these apes probably depended primarily on fruits found in the scattered clusters of trees, although they also may have picked the more sparsely distributed fruit and other plant food sources of the open savanna as they moved between the patches of forest. Eventually, however, the move from forest to savanna led to both anatomical and cultural adaptations that marked the emergence of the first hominids. The development of bipedal posture allowed these first hominids to range widely over the savanna and freed their hands to employ tools in their quest for food. Gathering—the obtaining of plant food with tools and carrying it for subsequent use—rather than hunting was almost certainly the crucial innovation that allowed these first hominids to fully adapt to this new grasslands environment where fruit and other foods were so much more widely scattered and less accessible than in the rain forest environment.

Gathering was probably more characteristic of females than males, for females had the additional burden of providing for the young—in utero, while nursing, and while still dependent although weaned. It is interesting that, although tool use in obtaining food is neither widespread nor regular among chimpanzees, females do use tools—such as stems to dip into termite nests or sticks to crack nuts—more often than males. In fact the skill female chimpanzees show in using tools and in passing their techniques on to their daughters suggests that among the early hominids tool-assisted gathering may have been a preeminently female adaptation, which spread rapidly as young females observed the new behaviors, imitated them, and were aided and rewarded by their mothers.

Hominid Expansion: *Australopithecus, Homo Erectus,* and *Homo Sapiens*

Using their bipedal posture and food gathering skills, the first hominids, members of the *Australopithecus* genus, spread slowly through eastern then southern Africa during the period from roughly 4 to 2 Mya. Between 1.5 and 1 Mya, members of the successor hominid species, *Homo erectus,* spread at a much more rapid rate through Africa and favorable areas of Europe and Asia, including Indonesia.

One feature that may have facilitated this further movement over much of the Old World was the development of better containers for carrying food, infants, and tools. Bone slice marks found on animal fossils in Africa dating from about 2 to 1.5 Mya and the much later slice marks on bones found in the Americas and known to be from skinning animals here (Great Plains bison, Eastern U.S. deer, and llama bones in Peru) are suggestively similar. Animal skins may well have been used to make the first good containers. Some 2 to 1.5 Mya women may have begun skinning animals to make better containers for carrying gathered food and babies. Women may have thereby invented a container that made possible a new spurt of human migration.

These marks on animal bones in Africa correspond in time to the early *Homo* finds there. It is *Homo erectus,* probably with containers far better than those of early *Australopithecus,* who spread out from Africa to Indonesia and China, then on to Europe as well. Their movement appears to have been accompanied by construction of more complex stone tools and the shift from predation without tools to hunting with tools. It is also with later *Homo erectus* in China that the first verified instance of fire use occurs.

As *Homo erectus* evolved into *Homo sapiens,* the migrations continued. Areas apparently untouched by *Homo erectus* were entered: the New World, Australia, and the far North. These initial worldwide movements by *Homo erectus* and then *Homo sapiens* brought gathering and hunting humans into most known regions of this planet. This expansion of gatherers and hunters over the globe was a movement of small groups of people, people who maintained small societies even after settling in a new area. On the whole, social organization did not need to be particularly complex for these small-scale societies. For the gatherers and hunters still living today, bands seldom number more than 100 individuals and usually average a good deal less. Kinship organization exists for all, but again is not complex. Ties are often loose and flexible, largely bilateral, with some patrilineally organized groups. Political organization per se is rare, and nation-states do not exist. Conflict takes only small-scale and personalistic forms as in fights and feuds.

After this spread over the world by gathering and hunting peoples, first horticulture, then agriculture were invented in several spots in both the Old and New World; in different areas different grains and animals were domesticated. The concept of plant and animal domestication for food and related skills spread rapidly over much of the world. Subsequent movements occurred: first of horticulturists, "hoe agriculturists" or shifting cultivators still living in fairly small-scale societies and often still doing some gathering and hunting as well, then of settled agriculturists who used more complex tools such as plows and who thereby could feed ever larger societies, living in larger and larger settled communities.

There began to be a few ecological difficulties; populations grew considerably; social structures and cultures diversified further; political units and small-scale warfare were invented. Even so, this expansion and further diversification of most peoples left the world reasonably undamaged. The continued human expansion on the globe also left its peoples using a wide range of subsistence forms, which apparently had been developed sequentially: gathering-hunting peoples in many warm areas, hunters and fishers in the very cold far north, slash and burn horticulturalists in many parts of the Old and New World, settled agriculturalists with especially growing populations in parts of the Mideast, Asia, Southeast Asia, the Pacific, and the New World. Some agricultural crops—major starches that could be cultivated in large amounts like wheat, oats, barley, rice, and corn and root crops like potatoes and yams—made possible far greater population concentrations in some areas than ever had been feasible before. In some regions—mostly among the gatherers and hunters living in very small-scale groups—specialized political forms still were not needed or used; class structure did not exist; both sexes were very involved in subsistence activities and often were quite egalitarian. Organized warfare, guided by governments and using specialized armies and specialized technology, did not occur in these many very small societies without agriculture or governments. It was the more settled, somewhat larger-scale societies that developed tribal units, states and nations—with larger and larger villages, towns, and cities—which came to have various types of specialized political structures and military forms.

At present human destructive technological capacities have outpaced the various social organizational forms that exist on Earth. For *Homo sapiens* to survive will require further social innovations to bring social behavior up to date with our new ability to be thoroughly destructive. One invention—warfare—has been developed to its carrying capacity.

Migration in Recent Human History: Colonialism as Contrasted to Minangkabau Merantau

European Colonial Expansion

Once humans had already settled the entire globe, further secondary movements outward began, including the one we are all so familiar with—the expansion over the globe of Europeans during the past several hundred years. Explorers set out to find new trade routes and new lands to exploit, missionaries followed to convert the heathen to Christianity, soldiers and officials to administer newly founded trading centers and colonies, and settlers anxious to build a new life overseas.

Francis L. K. Hsu has characterized the beliefs these Europeans carried outwards with them as "positive ethnocentrism"; he contrasts this with the "neutral ethnocentrism" which he feels is common to many other societies.

People with neutral ethnocentrism tend as much as those with positive ethnocentrism to see one's in-group as always right and all out-groups as wrong whenever they differ. But neutral ethnocentrists tend to have no desire to change the ways of those whom they see as inferior or wrong . . . proselytization to let the only truth prevail everywhere was confined . . . to people with positive ethnocentrism. That is why all the world's missionaries and missionary movements were and are of Western, and secondarily Arab, origin. On the other hand, even when the Chinese empire extended far and wide under Han, T'ang, Ming, no Chinese court had ever attempted to spread Confucianism or any other form of Chinese beliefs or ethics, and no individual Chinese ever received a call . . . to do the same.

This positive ethnocentrism characterized Europe's secondary movement outward and was a feature of Europe's colonial-type reexpansion over the past several hundred years. Although discovery and establishment of trade routes and missionization were what initially characterized Europe's secondary outward movement, governments soon went on to use their armies to take over political as well as market controls, often even to invade whole regions and to occupy the lands themselves. However crude this was, Europe got away with it; but not for long, for most colonized countries have long since sent the occupying Europeans home.

Japan tried a similar military-type reexpansion to resettle and control a wide part of the Pacific region, including Indonesia. But Japan got sent home even sooner than Europe. Subsequently, Japan appears to have applied most of its energies to export production and sales. Looking at its dominant position in the world economy, we may conclude that this was a highly beneficial move.

Minangkabau merantau

In contrast to the colonially oriented Europeans, and the militaristic Japanese, there are various peoples who appear to have been able to trade with, or migrate to, other parts of the world without destructiveness to either other peoples or themselves. They are people who have built a capacity to work in other areas into their cultures, adapting to other cultures rather than trying to convert other peoples to their own culture or religion. They have done so largely on an individual, family, and business basis, rather than through government or military channels. One example is the Minangkabau people of West Sumatra, Indonesia, who migrate from their homeland in accordance with their well-known custom of *merantau*. *Merantau* means "going out," often to trade or study. While gone from the homeland, people who merantau usually try to send help back home. Often, although not always, they temporarily return home at various times and/or return to live in the homeland late in life.

The Minangkabau are a matrilineal people. The matrilineage is a genealogically related group whose ties are through women and whose use rights to property are held by the women and pass through them. Matrilineages, or "wombs," are linked into even larger groups called *suku*, which contain male matrilineal figures with political roles that are significant not only within the suku but also act as representatives of the suku in the village council. In a village, homes are known by women's names. The kin group or family living in a house is a matrilocal, uxorilocal extended family: Three generations of women are common, and husbands are invited to live in their wives' kin homes. The extended family tends to focus about one of the mothers, a relatively senior woman. Sons and brothers retain close relationships to their matrilineal kin, especially to their sisters and mothers, and are helpful in terms of money, agricultural assistance, ritual, and a variety of kin activities, although men do live with their wives and much of the day-to-day help from men's earnings goes to their wives and children. The system has provided more respect, status, and involvement in kin group decision-making for senior women than common in traditional Western societies. There is also considerable economic security for women and young and marked freedom of movement and economic motivation for the men. Women control the homes and rice lands. It is the successful, helpful men who are most likely to be invited in as husbands. It is often men—though not only men—who engage in merantau. This custom of going out to other regions has made them aware of what is going on in the outside world, and they often are economically successful and socially comfortable in the regions to which they travel.

In their own homeland, the highlands of western Sumatra, Indonesia, the Minangkabau also tend to be internationally aware. Jewelry store owners in local towns know international gold prices, village men enjoy sitting in local coffee shops chatting and listening to national and international news on the radio, and teenage boys watch TV at night in the city square. Whereas youngsters twenty years ago still shouted Dutch phrases when white people walked by, today they say "Hello" or "Good morning" in English. Most Minangkabau are fluent in the national language, Indonesian, and know both a general form of Minangkabau and their village dialect.

The Minangkabau do not live only in their homeland of western Sumatra, and other parts of Indonesia. They also live across the Strait of Malacca, in Negri Sembilan, Malaysia, and trade with and in areas of Malaysia and Singapore, as well as in many parts of Sumatra, Java, and other regions of Indonesia.

Minangkabau merantau has taken several forms in Indonesia. One scholar, T. Kato, describes these forms as (1) village segmentation, (2) circulatory merantau, and (3) "Chinese merantau." Essentially, the first refers to a matrilineal segment moving from one overcrowded village to another village with more space; there they are adopted by a local matrilineage and receive use rights in land. The second form, termed circulatory merantau by Kato, refers to unmarried and married men earning their living largely outside their village or the Minangkabau homeland yet still retaining their primary home in their village, to which they return frequently. The third, Chinese merantau or *merantau Cino*, is a term used by the Minangkabau themselves to refer to a long-term migration of a kin unit to another area. Although Kato assumes these mobile families are always nuclear families made up of husbands and wives, I have also noted family women without husbands living away from their homelands—for example, a working mother with her child and her mother.

It is evident that merantau is a valued custom among the Minangkabau, a migratory custom that is known to have been carried out in several different ways over a long period. Merantau is supported by features of the Minangkabau social structure. In particular, the traditional homelands being in the hands of the women makes for relative economic security for women and children and allows marked freedom of movement for young men. But beyond that, interest in national and international news, fluency in the national language, Indonesian; interest in at least some learning of other languages and customs; and a relative cross-cultural adeptness all provide a cultural foundation for effective merantau.

The Minangkabau are usually characterized as traders. Indeed, it is not hard to find a Minangkabau woman selling beautiful gold jewelry on the streets of Jakarta. I particularly like to eat in Minangkabau restaurants whether in Jakarta, Jogjakarta, or San Francisco. But the Minangkabau were among the first Indonesians to develop a widely read national literature during this century. Although the Minangkabau are a far smaller ethnic group than the Javanese, the first president and vice president of Indonesia after independence were a Javanese and Minangkabau, respectively. Here at University of California, Santa Cruz, I am never surprised by the visit of a Minangkabau student or professor or of someone outside academia entirely who is on a trip to the United States.

For the Minangkabau, various forms of merantau continue to function effectively. Even as ethnic pride is maintained and many traditional customs retained, their approach to other peoples seems to be more what Hsu described as "neutral ethnocentrism" than "positive ethnocentrism."

Space Migration

Concepts similar to those supporting Minangkabau merantau seem more likely to be effective for interstellar migration than the "positive ethnocentrist" ideas and conquest-type policies that marked so much of the last few centuries of European colonialism. Europe's quick push outward combined the missionizing wish to change or convert others with rather awkward efforts to find new trade goods by control of other governments. The Europe of that period was somewhat inexperienced in long-distance cross-cultural travel or trade and substituted control for finding effective ways to operate long-distance trade in the new areas. The direction of the destructiveness during those few centuries of European colonialism was from Western countries to other peoples and areas. However, I think we now realize that destructiveness *from* other regions is also quite possible. So if only for this practical reason, effective skills in communication that include both economic and other negotiating skills with new peoples in new areas seem important to develop.

We might ask, why worry about such things when going into space? Maybe there are no other intelligent beings out there. Perhaps not. But there will be soon as humans expand outward, and after a while those several "human" groups of human ancestry will no longer look alike and their cultures will become different in many ways. These differences will probably develop quite rapidly in space, where the probability of long sexual and cultural isolation is at least as great as it was during earlier earth migrations.

The Infrared Astronomical Satellite (IRAS), in its brief period of existence in 1982, detected a dust cloud—a possible indicator of asteroids or planets—around the nearby star Vega. Until there is another means to detect planets, all we have is that tantalizing bit of information. Maybe there are planets orbiting around a nearby star. Maybe there are many planets in the universe waiting to be discovered—planets that may already hold life and that may soon nurture further expansion of life from Earth.

Right now the people developing space technology largely come from "positive ethnocentrist" cultures. In the recent past this was the European style of meeting with other peoples; now it is largely the style of the United States and Russia. Compared with those we may come to meet in space, the other Earth people met in Europe's reexpansion still had fairly similar biological and cultural heritages but looked and acted a little different: disease, raids, warfare and occupation, colonization, then revolutions and independence resulted. This process is not inevitable; people do not always engage in it as part of intercultural contact. But such a process is clearly part of the Western cultural heritage. It would be a shame if our descendants accepted these occurrences as migration inevitabilities. Probably the most important way we can assist solar and interstellar processes of human migration is to do practical, simple things right from the beginning that take what we know of the past into account.

Nearby Space Expansion in Our Solar System

For industrial expansion into the nearby space of our Solar System, Richard Lee's suggestion of simple licensing managed by a UN office could set precedents for both avoiding the exportation of extremely destructive, ugly types of industries into solar work areas as well as for seeing to it that some of the profits are sent back to help out this planet. I would like to make the further suggestion that some taxes on solar industries be set aside for interstellar research and exploration.

A Space Language

Of importance in interstellar travel are: mutual learning and reciprocal economic interaction when in contact with other intelligent beings, whether they are descendants of Earth peoples or extraterrestrial beings, rather than expansion by control over other ships, settlements, or planets, and methods of contact utilizing cross-cultural skills.

Potentially at least as important as either of these is the development of an interstellar language: an easy-to-learn space language for use in encounters between groups. Here Indonesian provides a practical example—one that could be borrowed. It is the second language of the peoples of the fifth largest nation in the world—a nation of many separate islands and many diverse cultures and languages but having much sea and air contact. Their national language, bahasa Indonesian, has root words that can readily be transformed into either verbs or nouns by the appropriate prefixes and suffixes. Therefore it is easy to learn, easy to speak, and practical.

Crew Members for Interstellar Flights

In preparing for interstellar exploration, a great deal can be done in crew selection. Anthropology and linguistics can probably contribute to effective interplanetary or intership contact, communication, and comprehension. Anthropologists and linguists as crew members could thereby contribute to intership or ship-planetary meetings either between Earth descendants or with other intelligent beings. Both anthropologists and linguists should be among the routine complement of any space crew.

As Birdsell suggests with regard to genetics, pick crew members with varying genetic backgrounds. This will also provide cultural richness, less boredom on long trips, and a range of interesting cooking knowledge. Similarly, crew members should regularly include women. A relatively equal number of skilled men and women will diminish intersexual conflict and intrasexual competition.

Some effort at teaching crew members about past and possible future problems related to positive ethnocentrism can be made through anthropology courses for the crew. Further cross-cultural skills for crew members can be developed through study abroad-type programs such as those now offered by many universities.

The traditional Western "positive ethnocentrist," racist, and sexist attitudes that have influenced past intercultural behavior of Europeans and others can be largely prevented for intraship, intership, and ship-planetary contacts by awareness of the possible problems, cross-cultural training for crews, and cross-cultural specialists among crew members.

Ship Facilities

Interstellar flights will be long and potentially boring. Along with the many flight skills, crew member hobbies will be important. Painting, music, weaving, jewelry making, sewing, and cooking good food from many parts of the world will contribute more than can be imagined now to maintaining interest on the long flight and to creating new planetary settlements. Small-scale ship gardening could be useful as a recreational activity as well as for long-term food provisioning. Sexual intercourse will, of course, provide one recreational activity. Birth facilities for women who want children should be readily available, as well as educational facilities and space for children's play. Similarly, sufficient contraceptives plus abortion facilities will be needed.

Interpersonal Encounters During Interstellar Migration

Knowledge of how other human societies—people living in small groups as well as those in large nations—live in our world can also provide us with ideas that may be helpful in thinking about the interactions of the crew itself during long trips within and especially between solar systems.

Until there are new innovations for much faster travel between stars, any interstellar travel will take very long periods of time. Space-traveling crews, separated from earth populations for extended periods, may well come to exhibit genetic differences, at least of a racial nature.

The societies today that have developed space travel innovations are large nation-states, relying economically on agriculture and industry. Most of their populations live in very large cities. Thus, the social background of the crews will, on the whole, be very different from their social life in space: The crews will be very small, highly skilled groups, often living together for more than one generation, and with no contact with other human groups for extended periods. The small-scale human groups that we know of are largely gathering-hunting peoples and shifting cultivators. Many are groups in which there is little political organization or hierarchy and to which both women and men make extremely important contributions to subsistence. Since these life ways are so different from our own recent history, it is essential that we be aware that such features can and do occur in small-scale groups.

Conclusion

In summary, cross-cultural and evolutionary studies can offer ideas about both social interaction among space crew members and encounters between a crew and interstellar planetary peoples living on some far world.

Recent cross-cultural data and information about evolutionary history indicate that there is a very wide range of human potential—that we have used our wide-ranging capacities to construct a diversity of human cultures. In fact, our species relies on its many cultures for human ecological adaptation. It is culture, with its basis in brain growth and intelligence, that *is* the human adaptation. Since cultural features constantly change, both by accident and by choice, it is likely that space crews will change, not only physically but also in terms of values, knowledge, skills, preferences, foods, rituals, arts, modes of social interaction, ways of socialization, and types of communication. Of interest for the future will be not only the space crews' own interactions or their interactions with any extraterrestrial beings that may already exist but also the interactions of crew descendants with crew descendants or the interactions of crew descendants with descendants of the remaining earth-living humans. Will we develop skills for merantau without "positive ethnocentrism" or colonialism? If not, who will colonize whom?

References

Boesch, C., and H. Boesch. 1981. Sex Differences in the Use of Natural Hammers by Wild Chimpanzees: A Preliminary Report. *Journal of Human Evolution* 10, 7: 585-593.

Hsu, Francis L. K. 1979. The Cultural Problem of the Cultural Anthropologist. *American Anthropologist* 81, 3: 517-532.

Jerison, H. J. 1975. Fossil Evidence of the Evolution of the Human Brain. *Annual Review of Anthropology* 4: 27-58.

McGrew, William C. 1981. The Female Chimpanzee as a Human Evolutionary Prototype. In F. Dahlberg, ed., *Woman the Gatherer*. Yale University Press, New Haven. Pp. 35-73.

Riss, David C., and Curt D. Busse. 1977. Fifty-Day Observation of a Free-Ranging Adult Male Chimpanzee. *Folia Primatologia* 28: 283-297.

Sibley, Charles, and Jon E. Ahlquist. 1984. The Phylogeny of the Hominoid Primates as Indicated by DNA-DNA Hybridization. *Journal of Molecular Evolution* 20: 2-15.

Tanner, Nancy M. 1981. *On Becoming Human*. Cambridge University Press, New York.

Wilson, Michael C. 1982. Cutmarks and Early Hominids: Evidence for Skinning. *Science* 298: 303.

Yunis, Jorge J., and Om Prakash. 1982. The Origin of Man: A Chromosomal Pictorial Legacy. *Science* 215: 1525-1529.

✳15

Douglas W. Schwartz

THE COLONIZING EXPERIENCE:
A CROSS-CULTURAL PERSPECTIVE

Predicting the social and cultural results of human colonization in outer space would be hazardous at best. Nevertheless, one means of providing a base from which such a prognostication could be attempted is to ascertain what has happened in the past to human colonies that have migrated to new surroundings. An examination of the anthropological and historical evidence of human migrations to determine the range of their results and the regularities that have emerged from them leads to some strong conclusions suggesting what might follow the establishment of human settlements beyond our planet.

Before we undertake any exercise regarding human colonies in outer space, however, it must first be acknowledged that interstellar human migration is possible. Then several additional assumptions must be accepted: First, humans sometime in the future will indeed want to leave Earth. Second, explorers will successfully locate suitable areas for human settlement, areas that possess the carrying capacity necessary to support a human colony. Third, when the explorers return to report their findings after decades of travel, conditions on Earth will be similar to those they left and there will still be an interest in migrating beyond this planet. Fourth, a powerful and cost-effective technology will be available to move out of Earth's atmosphere the materials and people necessary to establish a colony. Fifth, the migrating individuals will survive a trip to the new location, and the colony they establish will survive long enough for some social and cultural changes to occur.

If all of these conditions are met, it then becomes possible to ask what would happen to this colony. One means of understanding the potential results is presented by Ben Finney in chapter 9 in the examination of a critical case, the spread of humans over Polynesia. A second approach, which I use in this essay, is to explore the history of a broad range of human colonies that have migrated, analyzing the effect this movement exercised on the culture of the settlement and the sequence of events that occurred following resettlement.

Given the variety and uniqueness of human cultures, we may expect that a great part of any change occurring in space colonies will depend on the nature of the migrants' native culture as well as the nature and pressures of the new environment. But an analysis of past human migrations suggests there are some directions of change that recur with enough frequency to allow the results of a cross-cultural analysis of postmigration communities to be of real assistance in planning the colonization of space.

The Sample

This essay summarizes analyses of thirteen postmigration communities. Although few such studies have been undertaken, and none are available from industrialized societies, the insights provided by the inquiries that have been made into what happens to such communities and the regularities that result make this a useful heuristic exercise.

The cases used in this essay are from a variety of geographic regions and migration and cultural types. All the cultures examined are sedentary agricultural groups; they vary from primitive subsistence farmers to peasant agriculturalists having commercial relationships with urban areas. The groups used in the analysis, their location, the nature of their culture, the reason for their movement, and the type of migration involved are outlined below.

Migration within a region: Hopi Indians who moved from Oraibi to Moenkopi in northern Arizona, a range expansion because of economic pressure, overpopulation, and factionalism.

Migration between regions: Pioneering Americans who moved from the eastern side of the Appalachian Mountains to the Upper Cumberland River area of Kentucky and Tennessee in the late eighteenth century, searching for available farmland for settlement.

Modern American farmers who moved from the south plains of west Texas and Oklahoma to central New Mexico because of drought and economic depression toward available land and more favorable living conditions.

Neolithic Quechuan farmers in Peru, moving from inland to montaña locations over the past 100 years because of overpopulation.

Migration across a continent: Jewish farmers moving from Morocco to Israeli villages because of religious persecution.

Migration from a mountain valley to a coastal area: The Toba-Batak tropical farmers in Southeast Asia who moved from the coastal lowlands of East Sumatra toward available fertile land in the west because of overpopulation.

Migration from a continental location to islands: Tropical East Indian farmers who moved to the South Pacific Fiji Islands because of economic depression.

Migration between islands: Tropical Pacific farmers who because of overpopulation moved from minuscule atolls in Kiribati (Gilbert Islands) and Tuvalu (Ellice Islands) and from the tiny volcanic island of Tikopia to larger islands in the Solomon, Fiji, and Russell groups, respectively.

Migration across oceans: Basque pastoralists from Spain who moved to Idaho to improve economic conditions.

Northern Doukhobor farmers from Russia who moved to Canada because of a threat to core religious values.

Latvians moving to Brazil because of religious factionalism.

An analysis of these cases of migration suggests several themes that are frequently repeated, irrespective of the distance of movement, the motivation, or the type of base culture. These are summarized below as changes in total community configuration, technology, economy, social organization, and religion.

Total Community Configuration

Following a migration, communities may be said to proceed through three stages of organization: pioneering, consolidation, and

stabilization as the community reaches toward the successful establishment of its new settlement. Each phase has its own characteristics.
During the first two, factionalism can break the community apart and
lead to part of the group either moving back to the mother community
or establishing a new settlement, which would begin a new pioneering phase in another location. The sequence of new community
settlement is diagramed in figure 15.1.

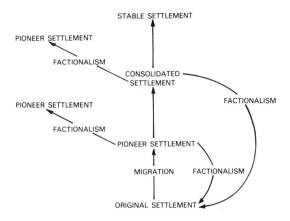

Fig. 15.1. Sequence of post-migration settlement types.

Pioneering Phase. The first phase of settlement usually covers the
initial two to four years following the colonization of a new area.
During this time the community directs its major efforts toward
physical survival. The first shelters are built and the first fields
planted, and every effort is made to establish a strong economic base.
This is always a difficult time marked by strong apprehension about
survival, an emotion expressed in the following passage in which
William Mortell describes the feelings of American pioneers.

> Fear and anxiety lay heavily on the minds of
> the people of the Upper Cumberland during the
> initial years of frontier settlement. These feel-
> ings were not manifested overtly; on the surface,
> settlers feared neither the known ene-
> mies—such as Indians, wild animals, floods,
> droughts, or blizzards—nor the unknown ene-
> mies, epitomized by death and supernatural
> creatures. Beliefs and feelings subtly couched in

oral traditions, however, reveal that fear was
indeed a compelling force.

In the earliest years of the new settlement, a high degree of group
solidarity is often necessary to offset harsh conditions, and a much
stronger feeling of ethnocentricity evolves than was present in the
premigration situation. This is particularly true if the community has
been established adjacent to an existing settlement and there is some
degree of conflict between the two, as occurred in Brazil among the
Varpa Latvian migrants, who developed feelings of ethnic superiority
over the local Brazilians. Such cases may ultimately result in reduced
contacts with or open hostility toward the local group. Frequently,
this period is also one of heightened group cooperation in construction and economic activities, with a high value placed on neighborli-
ness. In the first years the whole group may even live communally,
resulting in the development of cooperative work associations.

Although cooperation seems to be a characteristic of all pioneering
communities, if the families originate in different settlements with
contrasting or even slightly dissimilar values, the degree of cooperation is likely to be "spasmodic and uncertain," as Carl Dawson has
observed. Also, those individuals with less tolerance begin to chafe
under the hardships that are part of any pioneering venture and may
ultimately lose some of the feeling of group solidarity. The necessary
restrictions on individual action and the weight of authority that
accompany such periods of community development also serve to
discourage those who find themselves less able to adjust to this new
situation.

Because of these pressures in the earliest years of settlement, it is
common for one or two groups or families to break away from the
pioneer body and return to the mother body, where they know what to
expect. In Sumatra the early response of some Toba-Batak migrants
was discouragement and a return to the Meat Valley, their home area.
Some may migrate elsewhere with the hope of creating a situation
more closely related to what they envisioned their new life was going
to be.

Consolidation Phase. Generally, after the first or second successful harvest, more permanent shelters are constructed, a common
building may be added, and the settlement begins to move past the
feeling of impermanence that is always part of the pioneering stage.
This is when the community begins to consolidate, crystallize, and
formalize its social institutions and associations. The feeling of group

solidarity may continue through this phase, even after the needs of the initial settlement have been met.

A strong in-group feeling may take the direction of a rigid social order, where powerful sanctions are imposed against deviation from the developing social norms. This occurred among the Hopis who moved from Oraibi to Hoteville and dictated that close adherence to Hopi traditions was to be a paramount virtue. As a result, there developed a resistance to outside contact, cultural deviations, and change.

As the community evolves its new norms, there is often a strong attempt to retain the old ways even when change is necessary for survival. In the Tikopian colony, for example, deviations from the norm were explained as "temporary adjustments," implying the intention to return to the "real" ways at a later time. This is most frequently the case when a particular religious sect establishes a new community. Here enthusiasm for an ideal way of life will unite the members, no matter how different their original cultures. In time the purposes of the sect become so strict that members' lives may be defined in the most specific ways and, as Dawson writes, "even in their most intimate affairs." An excellent example of this trend can be seen in the case of the Mennonites, whose church precepts were rigidly enforced.

In other cases, especially those in which the reasons and conditions of movement were not as ideologically based, as with the Kiribati migrants to the Solomons, a strong feeling begins to emerge that the people need no longer be bound by norms dictated during the pioneering phase, and consequently they become receptive to culture change at an earlier time.

But just as there are individuals who cannot completely conform as the cultural pressures become more rigid during the pioneering phase, there are also those who feel the need to break away during the consolidation phase. It would almost appear that each stage taps a different level of individualism. Those with the strongest need for independence break earliest, whereas others are able to last longer, and a final group can adapt to the changed conditions and remain with the new settlement. This raises the interesting issue of the personality of the migrant. How do migrants differ from the stay-at-homes? If they are a special type, how do they survive once they are forced to settle into the new community? Is an early postmigration community composed of a higher percentage of malcontents than a settled community, since the former is made up of first-generation pioneers?

These are questions beyond the scope of this paper, but they have a real bearing on the potential success of any space colony. Factions that arise during the consolidation phase may or may not result in individuals leaving the community. Some of the disaffected may stay within the settlement but be identified as a separate group. Thus, the migration to New Mexico of Texas and Oklahoma farmers, whose cultural background stressed strong individualism, was characterized by factional splits that resulted in a proliferation of churches.

Other effects of factionalism during the period of consolidation include a greater number of political units, which serve to formalize differences; a dispersed settlement pattern; and the strengthening of kin ties as a means of reducing differences. As these various ways develop for dealing with disagreement within the community, eventually the direct effects of the migration pass and the group's economic problems begin to be solved. The community settles down to develop along lines not directly related to the resettlement and enters a phase of stabilization.

Technological and Economic Change

The amount of economic and technological change resulting from a migration seems to depend to a large extent on the differences between the original and the new physical surroundings. When a move takes place within the same environment, the traditional economy usually continues. If, however, a community moves into a new environment, technological and economic changes inevitably result. In the New Mexico community, a shift occurred from cotton and wheat farming to the cultivation of pinto beans. Piñon nut collecting was added, based on techniques learned from the neighboring Navajos. This illustrates a fairly strong pattern of technical change; namely, that new tools or techniques are generally borrowed from neighboring groups, even in situations where there tends to be little social interaction between the migrating group and the local settlement.

Further examples of this process can be seen among the Quechuas of Peru, who adjusted from a highland farming economy to one of tropical agriculture in the montaña region. Their new farming techniques were learned from montaña haciendas, where many worked during the first year or two after their migration. Agriculture and building techniques were also borrowed from local Indians. The Toba-Bataks, who moved from western Sumatra into a new environment on the eastern part of the island, were able to continue their traditional rice cultivation even though they were required to make changes in techniques to adapt to the leveler and more extensive plots that were now available to them.

Following a migration there may be a loss or decline of some nonutilitarian crafts. For example, the Toba-Batak women stopped weaving after their migration to East Sumatra because time and traditional materials were both lacking. Moreover, their new houses did not display traditional designs. Among the cases examined, however, there was not sufficient regularity to allow construction of a general pattern.

Social Effects of Migration

Alterations in social stratification, social organization, and authority patterns were present in many of the newly established communities reviewed. These changes occurred both singly and in combination, but their overriding direction seemed to reflect the emphasis of the base culture rather than being an effect of the migration. The range of each type of change is summarized below.

Social stratification. Migrations may initially have an equalizing tendency within a society. In the new Canadian settlements of the Doukhobors, for example, there was an early emphasis on socioeconomic equality among the settlers; and in the Indian communities of Fiji, emphasis on caste differences even decreased. This same effect can be seen at the Oren settlement in Israel, where a leveling of social distinctions occurred.

When social distinctions are strongly developed before migration and when less social displacement takes place afterward, the social class structure of the home area may be continued in certain groups. This occurred among Toba-Batak migrants and among the Varpa Latvians in Brazil. In the migrant community from Tikopia, the premigration class system was maintained; and although class differences were being obscured, indications were that they would be reinstated once the community was settled.

Changes in social units. Units of social interaction may be modified as a result of migration, although it is difficult again to make any generalization about what type of unit may become significant. The nuclear family household may become more important as a socioeconomic unit, with larger groupings decreasing in significance, as in the case of the Toba-Bataks; or extended kin relationships may become a primary means of widening obligations and responsibilities in a new settlement, as with the Fiji Indians, the Tikopians, and the Moroccans at Oren. A social unit with broad membership, such as a church, may become influential in integrating the new settlement, as happened among the Varpa Latvians and the Fiji Indians.

Traditional forms of social interaction, however, may also continue to hold the same significance as before the migration. This is generally found in postmigration communities where there was a short-distance move of fairly homogeneous populations within the same ecological niche, such as the Toba-Bataks' migration within the Meat Valley, the Mormon villages, and the Mennonite and Kiribati communities.

Authority. Both strong and weak authority patterns are found in postmigration communities, and the key factor is usually the nature of the pattern in the originating culture. A strong authority tends to be present in new settlements that result from well-organized migrations composed of cohesive social groups in which the traditional authority patterns and institutions are transplanted intact. This was the case with the Hopis who moved to Hoteville and with the Toba-Batak expansion within the Meat Valley. Strong authority patterns are also the result of the careful planning of a religious organization with strong central authority, as among the Mennonites, the French Canadians, and the German Catholics in Canada.

Weak authority patterns may be found in new settlements resulting from the migration of families or groups that were not previously living in the same community, as among the Toba-Bataks moving into East Sumatra and the Moroccan settlers at Oren. Among the Tuvalu Islanders on Fiji, who had migrated sporadically from the same community primarily to satisfy personal ambitions, authority patterns were also weak.

Again, the pioneer personality emerges as an issue. For if space travelers by their nature are individualists, then their relationship to authority may be a difficult one. If this problem arose spontaneously, an extremely divisive dynamic could be created within a postmigration community. It would seem much more productive to have a strong authority pattern well developed and accepted prior to the migration.

Religion

Three patterns relating to religion emerged from the survey of postmigrating communities. However, either the sample was not large enough to provide adequate data for predicting just when the three changes identified would occur or there are too many variables at work to permit such a prediction. The patterns are:

1. During the early years following a migration, a simplification of the religious system may take place.

2. Religion may become more important following a migration.

3. Religion may serve as a vehicle for factionalism after the initial period of settlement.

Simplification. Under the rigorous conditions of life in a new area, religion is likely to be affected by a loss of some ceremonies, a failure to replace some ceremonial objects, a general reduction in complexity, and a lack of personnel for ceremonial offices. After twenty-seven years of settlement at the new community at Bakavi, Mischa Titiev found, the Hopi ceremonial system was still fragmentary, with a paucity of traditional officers and paraphernalia. Although religion was important after the movement of the Toba-Bataks to East Sumatra, certain ceremonies, such as those associated with house building and family burial, declined. In the Indian communities on Fiji, religious gatherings were important occasions for social purposes, but there was a decrease in ritual.

Increasing importance of religion. In some postmigration communities, religion may become an important focus for unity. This was the case in the Mormon villages, where the church remained strong both in its ceremonial role and as a binding element for the new communities. Of course, when a religious sect is the active organizer of a migration, as in the Doukhobor case, there is likely to be substantial religious participation from the early days of the settlement.

Religion as a vehicle for factionalism. The Varpa Latvian settlement is an excellent example of a rapid split over the interpretations of the tenets of the faith occurring soon after the period of initial settlement. So many lay preachers developed as a result of this factionalism that the community could not support them as full-time specialists. Among the Fiji Islanders there was a proliferation of religious sects and associations, as was the case also in the Homestead and Toba-Batak communities. In the Tuvalu community of Kioa in Fiji, the church opened up a new channel through which the older and more conservative members of the group could assert themselves in the face of culture change after access to the traditional ways of asserting authority became closed off to them.

Conclusion

What might be learned from these cases of postmigration communities that could be applied to planning the human colonization of space? What variables appear to be of major importance in determining the direction a culture might take following its transplantation to a new location? Abstracting from the examples used in this study, five groups of critical variables relating to migration emerge: motivation, environment, structure, process, and culture.

Motivation. The purpose behind a migration in many cases directly affects the direction that change in the postmigration settlement will take. This is especially true when the move is prompted by personal gain, social or religious persecution, or economic need. Each of these motivations tends to force certain outcomes mainly because the types of individuals who are impelled to move for each of these reasons are likely to act in quite different ways once they reach their new homes. Those seeking personal gain will be strongly individualistic; people fleeing persecution are more likely to be part of a strongly cohesive group; and migrants in economic need will concentrate their activities, at least initially, on physical survival.

Environment. Once a decision has been made to migrate, the degree of difference between the origin and the destination is critical in determining the nature of resultant cultural change. If the movement is only a range expansion away from the home territory but still within the same ecological niche, there will be little social, economic, or emotional displacement. However, if the migrants move to an area some distance from their home community, with an environment quite different from that they have known, the cultural changes are likely to be massive. This would be the case with a space colony.

Migrants to an area with no immediate neighbors will not need to be as cohesive a group as would those faced with hostile neighbors. However, an existing population in a newly settled region would be a source of technological and economic ideas useful for survival.

Structure. Once the group has decided to move and the new location has been selected, then the structure of the group that will move becomes a major consideration. Are the migrants to be from a single community with a similar subculture or from different backgrounds? If they are chosen from the same subculture to foster long-term harmony, will they suffer from a smaller gene pool to the detriment of their long-term physical survival?

To what extent would the group be organized before the move? Would there be a loose focus, as random family members leave to search for new land in the same general area? Or would there be a strong focus and a tightly knit society, perhaps centered on a religion? *Process.* One of the factors determining the direction of cultural change will be the success of the settlement in the new environment. How long it takes to achieve some kind of economic stability will have a great deal to do with the extent of factionalism or group cooperation. A series of difficult years could result in a rapid simplification of ritual and of nonutilitarian crafts. This processual filter is a totally unpredictable variable but a critical one in determining the extent and direction of the change that will occur in the people's way of life.

Culture. Finally, while all these variables are at work, the most important factor is still the nature of the basic culture. The value system and its strength, the migrants' orientation toward change and individuality, how malcontents are to be dealt with, and a host of similar abstract issues may have more effect on the resulting success of the new settlement than all of the other determinants combined.

The two most important variables emerging from this review that can be controlled to some extent before a migration and that bear on its ultimate outcome are the interrelated factors of group cohesiveness and degree of individualism. It would appear that, all other things being equal, a successful migration depends upon a strong group orientation. Yet the creative loner may be just the type of personality that is necessary to produce the innovations and changes required once the new community is faced with the rigors of life in *terra nova*. So perhaps to the list of variables that must be considered in constructing a model for postmigration success should be added personality.

This paper is a further development of a chapter published in 1970 as part of a School of American Research advanced seminar.

References

Auguelli, John P. 1958. The Latvians of Varpa. *Geographic Review* 48: 365-397.

Cunningham, Clark E. 1958. *The Postwar Migration of the Toba-Bataks to East Sumatra*. Yale University Southeast Asia Studies. Yale University Press, New Haven.

Dawson, Carl A. 1936. *Group Settlement: Ethnic Communities in Western Canada*. Macmillan of Canada, Toronto.

Edlefsen, John B. 1959. Enclavement among Southwest Idaho Basques. *Social Forces* 29, 2:155-158.

Knudson, Kenneth E. 1964. Titiana: A Gilbertese Community in the Solomon Islands. *Field Studies in the Project for the Comparative Study of Culture Change and Stability in Displaced Communities in the Pacific*, Homer G. Barnett, director. University of Oregon, Eugene.

Larson, Eric H. 1966. Nukufero: A Tikopian Colony in the Russell Islands. *Field Studies in the Project for the Comparative Study of Cultural Change and Stability in Displaced Communities in the Pacific*, Homer G. Barnett, director. University of Oregon, Eugene.

Mortell, W. L. 1983. Don't Go Up Kettle Creek. In *Verbal Legacy of the Upper Cumberland*. University of Tennessee Press, Knoxville.

Schwartz, Douglas W. 1970. The Post-Migration Community: A Base for Archaeological Inference. In William Longacre, ed. *Reconstructing Prehistoric Pueblo Societies*. A School of American Research Advanced Seminar Book. University of New Mexico Press, Albuquerque.

Steward, N. R. 1965. Migration and Settlement in the Peruvian Montaña. *Geographical Review* 55:143-157.

Titiev, Mischa. 1944. Old Oraibi: A Study of the Hopi Indians of Third Mesa. *Papers of the Peabody Museum of American Archaeology and Ethnology* 22, 1.

Vogt, Evon A. 1955. *Modern Homesteaders*. Belknap Press/Harvard University Press, Cambridge.

Weingrond, Alex. 1962. Reciprocal Change: A Case Study of a Moroccan Immigrant Village in Israel. *American Anthropologist* 64:115-131.

White, G. M. 1965. *Kioa: An Ellice Community in Fiji*. Department of Anthropology, University of Oregon, Eugene.

"For nearly three centuries the dominant fact in American life has been expansion. With the settlement of the Pacific coast and the occupation of free lands, this movement has come to a check. That these energies of expansion will no longer operate would be a rash prediction."

—*Frederick Jackson Turner*

"I dislike arguments of any kind. They are always vulgar, and often convincing."

—*Oscar Wilde*

*16

Edward Regis Jr.

THE MORAL STATUS OF MULTIGENERATIONAL INTERSTELLAR EXPLORATION

A multigenerational interstellar voyage is manned space flight that would take so long to complete that the individuals who boarded the ship when it departed would not be alive when it reached its destination. The trip would be so protracted that it could be completed only by the descendants, perhaps the very remote descendants, of the original crew. The variety of scientific, technological, economic, and ethical problems in the way of such a trip, their number, depth, and difficulty, is staggering. In fact, the primary obstacle to our taking the idea of such a trip even semiseriously will be the inability to think that the trip could ever be possible in principle. For we would have to imagine that it will someday be possible to construct a spacecraft vast enough to carry a population and life-support system the size of a small city, to accelerate this craft to a speed that would enable it to reach a habitable Solar System in a reasonably short span of generations, that the populations aboard could maintain themselves throughout all this time, both physically and psychologically, without any outside assistance and, finally, that there is some good reason for attempting this grand venture in the first place. Fanciful or not, the general design of such a craft, now often called a space ark, was first suggested by J. D. Bernal in 1929:

> Imagine a spherical shell ten miles or so in diameter, made of the lightest materials and mostly hollow. . . The great bulk of the structure would be made out of the substance of one or more smaller asteroids, rings of Saturn or other planetary detritus. . . The globe would fulfill all the functions by which our earth manages to support life.

This craft would contain living quarters, propulsion, navigation and control systems, and a closed, self-sufficient agriculture and waste-recycling system capable of sustaining several generations of inhabitants. The usual propulsion methods have been suggested for such a vehicle. Some examples are the fusion rocket, the nuclear pulse rocket, the Bussard interstellar ramjet, the laser-photon sail, and so forth. In addition, two exotic science fiction methods have been proposed: tachyon and antimatter propulsion. Assuming that such a craft could be built, manned, and accelerated to 1 to 2 percent the speed of light, then travel to the nearest stars, the Alpha Centauri complex, would require about 400 years of on-board travel time. Assuming also an average life span of 75 years and couples having exactly two children at about age 25, then fifteen generations would have to be born, reared, and educated on this flight.

But now that we know how to get there, we might ask: Why go? The usual motives offered are: to satisfy human curiosity, to escape the limitations and/or possible dangers of continued life on Earth, to contact other intelligent civilizations, to colonize other planets, to challenge an infinite frontier. But it is difficult to predict which, if any, of these motives might operate in future centuries to make such an expedition desirable or even necessary. The history of space exploration to date suggests that more than such noble motives are required to move humans off the planet. The United States went to the moon in July 1969, not primarily to escape the bounds of Earth nor to satisfy a need to explore nor to venture out upon a new frontier. Our primary, stated reason for going was far baser: It was to outdo the Russians. What about the Russians themselves? Were they motivated by a thirst for knowledge and a desire to explore when they launched Sputnik in 1957? They probably were to some degree, but political incentives cannot be underestimated. Of the two Soviets most responsible for Earth's first artificial satellite, only one was a scientist, Sergei Korolev; the other was Nikita Khrushchev.

But even allowing for all the noble and ignoble motives one could imagine, it is hard to think of what could justify the expenditure of energy required for interstellar voyaging. In a recent paper Frank Drake has calculated that the energy needed to send a colony of 100 people over a 3 parsec distance in 100 years is equivalent to "the total energy necessary to meet all the energy requirements of a major country, such as the United States, for a period of time of *hundreds of years*." He adds: "It is very difficult to believe that any intelligent group of people, or any major government, would ever consider this a cost-effective approach to any problem." The "Where Are They?" question becomes quite relevant here, for everything that makes interstellar travel and colonization unlikely for *them*, civilizations far in advance of our own, reduce by an even greater degree the likelihood that we will ever be able to achieve it ourselves.

At worst, in view of its difficulties, a trip such as we are imagining is an exercise in fantasy. At best, what can be said for it is that it does not seem to violate any known laws of nature. But for the sake of getting on with our work, let us grant that at some time in the future, all these conditions will be fulfilled. The scientific and technological problems will be overcome, there will be good reasons for going, and the required energy will be available. What then?

The moral problems must be considered. For example, How ought the journey be financed? Can the entire world's, or the Solar System's, population be taxed for this purpose, for a journey from which *they* will not benefit? What criteria ought to govern crew selection? Must every race or nation be represented? What would be the legal status of the starship? Would it be a nation unto itself, floating through space? How ought it to be governed? Democratically? Tyrannically? Should new generations be told of Earth and told they are on a "mission" whether they like it or not? Could mind control of some kind be practiced to keep the mission going? Would it be morally legitimate to colonize a world with life, perhaps intelligent life, already on it? Do humans in fact have the moral right to use and exploit the Galaxy for their own selfish purposes? With respect to the latter question, I note that at least one writer, David Thompson, has bemoaned the lack of any "comprehensive plan for conservation of the resources of outer space." He maintains that such plans "must be laid now, before exploitation begins and vested economic interests develop."

It would be a mistake in the face of such concerns to think that the only obstacles to interstellar migration are technological or scientific.

There are serious ethical and political problems that must also be addressed for interstellar travel ever to become a reality.

The specific moral problem that I propose to consider is whether a multigenerational interstellar flight could be morally permissible. The main reason for thinking it could not be is that the flight will confine many generations to a small, artificial environment for their entire lifetimes, whereas their consent to such confinement has been neither asked for nor given. This is equivalent to locking up large numbers of people and throwing away the key, a procedure that appears to violate human rights in a fundamental way. On a very basic, intuitive level, such a flight seems wrong, a case of interstellar kidnapping, involuntary exile, or repeated child abuse. Does any generation have the moral right to place its descendants in such a position?

One might think that by the time such a voyage were possible, humanity already would have moved off the face of the planet and be living in various Solar System habitats, so that the idea of wrenching generations from their natural home, Earth, would not arise. But the moral problem I am posing may pertain even to colonies within the Solar System. Perhaps for one reason or another we have no right to confine our children even to them. I propose to address both these problems at once, by using the concept of a multigenerational flight as a test case. If *it* will be morally permissible, then so will any other form of space colony, whether within the Solar System or outside it.

I begin with what seems to be the most obvious objection to a multigenerational flight: that it violates the rights of the new generations, since they are placed for their entire lives into conditions they have not chosen and from which they cannot escape. At first glance this point seems so decisive that it appears no more need be said. But a moment's consideration will show that precisely the same is true of any new generation on Earth, for all of us are placed at birth into conditions (the state of the world at the time of our births) that we have not consented to and from which we cannot escape. Indeed, in this most fundamental of ways, Earth and the spacecraft are exactly on a par. But if this were true and were all there was to be said, then life aboard the spacecraft would be no more or less morally permissible than life on Earth, a conclusion at odds with our initial strong intuition. There must be some difference between them that we are overlooking. What is it?

The two most obvious differences are (1) the spacecraft just is not the Earth and (2) the spacecraft's living conditions will be far more restrictive and confining than Earth's. Someone could contend that either of these differences is enough to make the voyage ethically impermissible. The argument would be that since humans evolved here, any human being has a kind of natural right to be on Earth; the interstellar flight violates that right and is hence immoral. Or the argument would be that a spaceship is no place for a human being to spend his or her entire life: it is constricting, artificial, and sterile.

But despite appearances, neither of these arguments is any good. To start with, is it in fact true that everyone (or anyone) has a right to be on Earth as opposed to elsewhere? As far as I know the question has never arisen before, and for the sufficient reason that there was never any possibility of being anywhere else. (I ignore whether being sent to heaven or hell violates your right to be on Earth.) Nevertheless, it is not immediately clear what the answer to the question is. In virtue of what feature about Earth or about its inhabitants could human beings be said to have a right to live here, supposing it would be possible for them to live elsewhere?

As with many problems associated with situations that have not arisen before, one way of approaching a solution is by the principle of closest analogy, as is often done in law. But what analogy could we appeal to when the question is: Is there a human right to be on Earth? No similar questions spring immediately to mind. The closest analogy might be to ask whether there is any general human right to own or live on any minimum amount of Earth or on a particular plot or parcel of Earthly land. If there were a general right to possess some particular part of Earth, then there would be a derivative general right to be on Earth, a right that a multigenerational flight would, of course, violate.

But there are really two questions here. One is whether there is a right to any given minimum amount of anything; the other is, if so, is that thing some part of Earth? So how could we establish a general right to any minimum amount of anything? It seems we would have to start by presupposing a general right to own property. If there were such a right (which is doubted by some thinkers), there would not follow from this alone a right to own any minimum amount of property. Rights to specific amounts, pieces, or parcels of property arise, according to theories of John Locke, Robert Nozick, and others, only through original acquisition or transfer. In that case, one would not come into the world owning any property at all, except perhaps oneself.

Other theories, however, claim that one comes into the world bearing a right to life and to a certain minimum amount of utility, welfare, or well-being. Even on this theory no right to specifically *Earthly* property follows. For if the requisite amount of utility, welfare, or well-being could be provided on other planets, and there is no reason to think it could not be, then the right to well-being could be fulfilled elsewhere than on Earth. This means that no right to be on Earth follows from a right to well-being alone.

Perhaps some other theory of rights, such as an egalitarian right to an equal share of Earth's resources, would provide the basis for a general right to be on Earth. For if, as egalitarians claim, all people have an equal right to all things or a right to an equal share of all things, then it looks as if people have equal rights to Earth's resources and hence a general right to be on Earth.

But this encounters several problems. Why stop with just Earth's resources? There are fabulous resources on the Moon, the asteroids, and the other planets and their satellites. Why does everyone not have an equal right to them as well? If they do not because such resources have not been developed, then it is hard to see how they could have (as egalitarians suppose) any right to the as yet undeveloped resources on Earth, for example, the oil still in the ground, the grain not yet harvested, or the labor not yet performed. But even if there were rights to equal shares of all these things, to the developed and the undeveloped resources of Earth, the Moon, and all the planets, there still would not follow rights to any particular bits of these, nor hence rights to be any one or more of those places. The most that would follow from a theory of strict egalitarianism would be a right to an equal share of the whole, whatever that may be taken to be. But one could exercise one's right to that share on many places other than Earth, just as one can, on Earth, exercise one's right to an object without that object's being in one's physical possession. For these reasons, the existence of locations other than Earth where rights to equal shares might be exercised prevents such rights from being rights to *be* on Earth.

That people do not in fact have any general rights to be on Earth may be seen from the following. Suppose that a planet were discovered or created out of asteroids just the same size, composition, and mass as Earth and with the same atmosphere and distance from a star such as the Sun. Suppose further that the planet were terraformed to resemble Earth in most but not all details: Call this planet "Doublearth." Now could children born there of parents who migrated to Doublearth claim that their rights are violated because they are not on Earth (the real Earth)? I do not see what could warrant such a claim. Suppose further that instead of being the same size as Earth, Doublearth were 10 or 20 percent smaller. Would the rights of settlers' descendants be violated in that case? If Doublearth were only 50 percent the size of Earth and quite different in composition and topography but fully habitable, would that violate anyone's rights? Not obviously. It seems then that descendants of Earthlings have no more right to be on Earth than descendants of Europeans who migrated to America have to be in Europe.

You may think this is too hasty. For to see what the denial of a right to be on Earth amounts to, we must consider some concretes. It will mean that people do not have rights to see and experience such things as the Grand Canyon, a sunset, a rainbow, a night at the Metropolitan Opera, or a week in Paris. Can it really be that people do not have rights to be where they can enjoy such things? Do you think you have such rights? Well, you don't!

If the Grand Canyon were made into a lake, if the residents of Paris all left taking their hotels, fine wine, and restaurants with them, if Metropolitan Opera singers unanimously took up other careers, these events would not appear to violate the rights of any other person. To affirm general human rights to things simply because they are always there, because you like them very much, or because you take them for granted is, in the absence of any stronger justification, sheer wishful thinking.

Is there a general right to be on Earth then? Although I do not claim to have proven there is not, I have given some reason to think there is no such right. The burden of proof remains on the affirmative, and until and unless that burden is fulfilled, there is no ground here for objecting to a multigenerational interstellar journey.

We return then to the other possible ground for thinking that the spaceflight would be morally impermissible, namely, that it would intolerably restrict human freedom. Although such restrictions are permissible if accepted voluntarily by the original crew, perhaps the crew could not force others against their will into such restrictive conditions. This means that the original crew could not morally procreate aboard the space ark, and so the trip could not be made.

But here a lot depends upon what conditions aboard the spacecraft will be like. The best analogy, one that has been suggested by some who propose the journey, is that of life aboard a huge ocean liner, one large enough to accommodate 10,000 people in relative comfort for their entire lives. But how to decide whether even this is an impermissibly constricting environment for a lifetime?

We can start by noting that procreating in some restrictive conditions is not usually regarded as immoral. East Berliners, for example, seem to have a perfect moral right to procreate even if their children were to be confined forever to that city. Nevertheless, there must be some conditions extreme enough to make procreation wrongful. Most people would probably agree that it would be wrong to procreate and raise children in a Nazi concentration camp, supposing that were even possible. In fact, it was not possible, even biologically, for in the Nazi camp sexual desire was extinguished for both men and women. Nonetheless, it is not wholly clear that procreation even in such extreme conditions would always be immoral. Childbearing has occurred in Russian forced labor camps; it is not clear that this is immoral. Slaves in the United States often bore children prolifically and it is not obvious that this was immoral either.

Contemporary philosophical discussions of children's rights to a minimum standard of living as against parents' rights to procreate even if they cannot guarantee such a standard do not throw much light on the problem. Onora O'Neill, for example, suggests that prospective parents have no right to procreate in situations "such as serious ill health or abysmal poverty." This would make the preponderance of historical and perhaps even contemporary procreation immoral, a conclusion probably not acceptable to many of those who procreated in those conditions, nor to their offspring. Another theorist, William Ruddick, is more liberal. He claims that a parent must foster life prospects that, if realized, would be acceptable to both parent and child. But this is subject to an obvious difficulty: How can anyone know, in advance of its even being conceived, what life prospects might or might not be acceptable to an unborn child?

But no matter what the truth is with respect to procreation in extreme conditions, it must be recognized that conditions aboard the spacecraft, at least as these are usually described by the proponents of interstellar travel, are not even remotely as constricting as those in which many people bear children today. Far from there being severe deprivation aboard ship, all the generations to be born there are to be well provided for throughout their life spans, something that could not be said for most of the children who have ever been born on Earth.

This covers only physical conditions. Perhaps the star voyagers may yet lack a measure of psychological or mental freedom.

Will there be enough people aboard for variety in human companionship, for adequate choice in selecting a romantic partner, or for a stimulating and well-developed culture? But however we may answer, we must understand that here we are no longer speaking of that to which anyone could justly claim a right. For although these conditions may be necessary for an ideally fulfilling life, it does not seem that people have rights to such lives nor therefore to the conditions that would make them possible. It does not seem to be true, for example, that children have rights to live in a city of a given size or larger or in one having a certain minimum level of cultural sophistication. As reference points, consider that the early colonial population of several American states was less than 10,000 each, less than the projected population of the starship. Many South Pacific islands even today have populations well under 10,000, some islands and atolls having relatively isolated communities numbering as low as 300 people. Finally, there are cases even in contemporary Western society of small populations living in isolated conditions for long stretches of time, for example, people living on remote farms in Norway, isolated from practically everyone else but other members of their family, and the Hutterites, a religious sect living in geographically and socially isolated collective agricultural colonies in northwest United States and Canada. Unless procreation in all these isolated Earthly circumstances is immoral, which I see no good reason to believe, we must conclude that procreation aboard the multigenerational spacecraft would not be wrongful either.

But perhaps there will be too many people aboard instead of too few. Perhaps the crew will suffer overcrowding and a lack of privacy and psychic space that will make the journey morally objectionable after all.

As before, everything will depend on the precise conditions to be found aboard the vessel. If its living conditions satisfy the minimal requirements for normal human living space, then the trip would be permissible on this criterion; if not, the trip would not be permissible. But do we even know what these requirements are? This is uncertain, but what we do have are many empirical studies of humans in isolated and confined environments, for example, those in Arctic and Antarctic scientific and military installations, nuclear submarines, undersea labs, and United States and Soviet orbiting laboratories. This research suggests that although human space requirements are not absolute, there are definite lower limits of volume, and especially of privacy, that must be respected if human life is to be bearable under conditions of long-term restriction. It follows then that so long as the requisite amount of living space and privacy were to exist aboard the space ark, there could be no objection here to procreation aboard it.

To summarize, a multigenerational interstellar expedition does not seem to be morally impermissible on any ground that we have considered. It does not appear to violate any right of which we are aware, not an alleged right to be on Earth, not children's rights, not rights to freedom, living space, or privacy.

Any residual aversion to the idea of a space ark probably stems from our imagining what we would be missing if we ourselves were to be on it for our lifetimes. For we would be denied all those things that are to be found only on Earth, things that we value deeply. Moreover, parents want their children to have as good, if not better, lives than they themselves had, something it is hard to think will be possible aboard a space ark no matter how large and well fitted out it is. It is understandable then that we tend to recoil from the thought of generations enclosed for all their lives in an artificial environment and without the pleasures of Earth that we so value. Anyone who enjoys driving across the Mojave Desert or skiing in the Rockies or watching the moon rise is going to be somewhat at a loss in a spacecraft. But against this must be placed the fact that these generations, because they will not have known Earth to begin with, will not know and hence not miss what they are missing. The spacecraft will be home to them, the only environment they know by direct acquaintance.

But what does this prove? Surely, it is not true in general that so long as people have not known anything else it is permissible to confine them to what they have known. This would legitimize straitjacketing a child for life, or keeping him enclosed in a closet, or the Chinese practice of binding the feet of female infants. But what is it that separates moral wrongs such as these from a moral right such as procreating when an East Berliner? The answer can only be that what separates them is the degree to which one's freedom of action is restricted. Whereas freedom of travel, speech, and emigration may be compromised in East Berlin, they are obviously far more seriously restricted by a straitjacket or a concentration camp. It is morally permissible to procreate in East Berlin because it is morally permissible to procreate in less than perfect or ideal conditions. If this were not so, then most of the people who were ever born on Earth would probably have been procreated immorally. But even if they have been immorally procreated, we have seen no reason to think that conditions on the spacecraft would remotely approach the state of deprivation to which most of humanity has been subject.

Against all of what has been said, however, it might be alleged that the expedition is wrong because its senders would be using others for their own purposes. They would be treating the crew as mere means to the senders' own ends instead of respecting them as ends in themselves. But the answer to this is that although the crew is being used for others' ends, it does not follow from this alone that the crew is not treated as ends in themselves as well. It is possible to use others as means while also treating them as ends in themselves; to do the latter we merely have to respect their rights. People have long used their children as means to their own ends and indeed still do so: A recent book on children's rights begins with the observation that "Children are the most valuable resource of mankind." But there is nothing wrong with considering children as resources provided that in using children as means to their own ends, parents violate none of their children's rights. Because, as we have argued, confinement to a space ark violates no rights, there could be nothing morally objectionable about the senders' use of the crew for the senders' own purposes.

It is true that in the designing, outfitting, and launching of a space ark, planners will be committing others to conditions not of their own choosing. But this, as Iain Nicolson says, is just what happens, and necessarily happens, on Earth: "The results of our deeds and misdeeds on this planet come home to roost for future generations; the crises we face on this planet today are—in part at least—due to the decisions or indecisions of our forefathers." A multigenerational interstellar expedition is no more and no less than a microcosm of human life on our own planet. Launching one is like beginning Genesis anew: It is to place human beings on another heavenly body for its inhabitants to make of what they will. To do this, to put humans on a microplanet upon which they can live, procreate, dispose of their affairs under their own governance, and die, is to bestow upon them conditions quite analogous to those on our own planet, conditions they may use for good or for ill, just as we use the planet Earth.

Despite all appearances, a multigenerational interstellar expedition does not differ in any relevantly fundamental way from ordinary human life on Earth. It is true enough that in its sheer vastness, beauty, and variety, Earth has incomparably more to offer than a space ark, but there is no one person on Earth who could partake of all the beauty and variety that our planet has to offer. Many, perhaps most people, whether through ignorance, necessity, or free choice, partake of an extremely small portion of it. Although a space ark may seem, when compared with Earth, to be a sterile, impoverished, and alien environment, it will be, we have argued, as much of an environment as anyone could have a natural right to and a better environment in many ways than a large number of people have ever enjoyed on Earth.

We conclude then that just as no rights of Earthlings are infringed by their not being included as passengers on a space ark leaving for the stars, no rights of star voyagers are infringed by their not living their lives on Earth. A multigenerational interstellar expedition is no more and no less morally permissible than the very existence of human life on our own planet.*

*Earlier versions of this paper were read at Howard University, the University of the District of Columbia, and at the Reason Foundation Summer Seminar, Santa Barbara, California. For their helpful suggestions I would like to thank J. C. Smith, Randall Dipert, Tibor Machan, and especially Charles Griswold. I am also indebted to Kerry Joels, National Air and Space; Jesco Von Puttkamer, NASA Headquarters; and Jane Riddle, NASA Goddard Space Flight Center, for their assistance.

References

Aiken, William, and Hugh La Follette, eds. 1980. *Whose Child?* Rowman and Littlefield, Totowa, NJ.

Bernal, J. D. 1929. *The World, the Flesh and the Devil.* Methuen, London.

Carroll, Verne, ed. 1975. *Pacific Atoll Populations.* University Press of Hawaii, Honolulu.

Cleaver, A. V. 1977. On the Realization of Projects: With Special Reference to O'Neill Space Colonies and the Like. *Journal of the British Interplanetary Society* 30:283-288.

Conquest, Robert, 1978. *Kolyma: The Arctic Death Camps.* Macmillan, London.

Des Pres, Terence. 1976. *The Survivor: An Anatomy of Life in the Death Camps.* Oxford University Press, New York.

Drake, Frank D. 1980. N is Neither Very Small Nor Very Large. In M. D. Papagiannis, ed., *Strategies for the Search for Life in the Universe.* D. Reidel, Dordrecht, Holland. Pp. 27-34.

McArthur, Norma. 1968. *Island Populations of the Pacific.* Australian National University Press, Canberra, and University Press of Hawaii, Honolulu.

Mallov, E., R. L. Forward, Z. Paprotny, and J. Lehman. 1980. Interstellar Travel and Communication: A Bibliography. *Journal of the British Interplanetary Society.* Vol. 33.

Mazlish, Bruce, ed. 1965. *The Railroad and the Space Program: An Exploration in Historical Analogy.* MIT University Press, Cambridge.

Nicolson, Iain. 1978. *The Road to the Stars.* William Morrow, New York.

Oberg, James E. 1981. *Red Star in Orbit.* Random House, New York.

O'Neill, Onora, and William Ruddick, eds., 1979. *Having Childen: Philosophical and Legal Reflections on Parenthood.* Oxford University Press, New York.

Paterson, Erik J. 1981. Space Settlements—The Medical Perspective. *Journal of the British Interplanetary Society* 34:429-434.

Richards, I. R. 1981. A Closed Ecosystem for Space Colonies. *Journal of the British Interplanetary Society* 34:392-399.

Ruddick, William. 1979. Parents and Life Prospects. In Onora O'Neill and William Ruddick, eds., *Having Children: Philosophical and Legal Reflections on Parenthood.* Oxford University Press, New York.

Thompson, David. 1978. Astro-Pollution. *CoEvolution Quarterly* (Summer): 35.

SECTION IV

Speciation

INTRODUCTION

There are limits to using past experiences of our species to think about the human future in space. In fact, if our descendants do succeed in scattering far and wide through the Galaxy, there will be no single future for humanity. Indeed, there will be not one humanity, but many. Even with warp drives, hyperspace travel, and other literary inventions for traveling faster than the speed of light, science fiction writers have immense difficulties keeping their galactic empires together. We make the assumption that for at least a very long time interstellar travel speeds will not exceed some minor fraction of the speed of light. Hence, distances of many light years between successively settled star systems (not to mention wandering interstellar comet communities) will divide our spacefaring descendants into myriad independent entities. With independence will inevitably come diversity, first cultural then biological.

Natura non facit saltum. Darwin rigorously adhered to this motto, "Nature does not make leaps," supposedly first penned by Linnaeus. This tenet of Darwinian evolution has recently been challenged by Stephen Jay Gould, Niles Eldredge, and other paleontologists who maintain that much of evolutionary change proceeds through a process they label "punctuated equilibria." They see the fossil record as one of long periods of stasis punctuated by short bursts of speciation, rather than the slow and gradual transformation of one species into another that Darwin imagined.

In our paper, "The Exploring Animal," in the first section of this book we proposed that dispersion into space will, by dividing humanity into innumerable small and reproductively isolated groups, lead to rapid speciation. In so arguing we are going against the opinion shared by many anthropologists that cultural evolution has virtually taken over from biological evolution, as well as the belief of Darwinian gradualists that only slow, incremental change will occur in our physical evolution. We even find ourselves going against the thinking of a pioneer of the punctuated equilibria concept, Niles Eldredge, who, writing in *The Myths of Human Evolution* (Columbia University Press, 1982) with anthropologist Ian Tattersal, asserts that the "possibilities of a speciation event within our lineage in the foreseeable future are two: slim and none."

Where we differ from Eldredge and Tattersal is in what is to be included in the "foreseeable future." They foresee the establishment of human settlements on Mars and elsewhere in the inner Solar System, but no farther. If that were to be the limit of human expansion they would undoubtedly be right, for unless our descendants do scatter far and wide beyond our Solar System it is difficult to foresee how speciation could occur, except as the cumulative result of hundreds of thousands, if not millions, of years of slow and gradual genetic change. However, at least two adherents of the punctuated change school do share our vision of rapid speciation through dispersion into space: Steven Stanley, who ends his book *The New Evolutionary Timetable* (Basic Books, 1981) with such a view, and James Valentine, whose cogently argued essay comes first in this section.

Valentine, drawing on the fossil record of his specialty, marine paleontology, suggests that evolutionary novelty appears most frequently and intensively where environments are open to colonization. He therefore proposes that if the Galaxy is really empty of competitors, space expansion will result in the origin not only of new hominid species but also of wholly new forms of intelligent life. Wisely, however, Valentine declines to predict what these new phyla, classes, and orders of space-adapted life might be like, other than to say that if we could contemplate them we would be truly astounded.

Whereas Valentine focuses primarily on naturally occurring processes in evolution, in the second paper astronomer Michael Hart boldly predicts that genetic engineering will be used in space to affect radical changes in body form and chemistry for better adaptation to new space environments and also for lengthening life spans to up to three millennia. Although living 3,000 years struck conference participants as both difficult to achieve and to endure (especially if

no cure for arthritis were found!), the inevitability of genetic engineering for better space adaptation seemed more plausible and intriguing. In fact, it fits in with Valentine's point that it will be those
forms with unusual talents for spacefaring whose descendants will
spread across the Galaxy. Hart, in arguing against the common thesis
that one day our descendants in space will outgrow our "adolescent"
penchant for expansion and settle down peacefully where they are,
proposes that there will always be a few aggressive groups that will
wish to continue the expansion. It will be they, he asserts, who will be
most tempted to employ genetic engineering to better adapt themselves for colonizing new space environments. Although one may
argue that Hart's scenario of the Galaxy rapidly filling up with
aggressive human descendant forms is exaggerated, it does make us
think twice about the nature of our urge to explore and its consequences.

✻17

James W. Valentine

THE ORIGINS OF EVOLUTIONARY NOVELTY
AND GALACTIC COLONIZATION

The time scales against which we contemplate the expansion of the human species out across the Galaxy are very long indeed when compared with those involved in practical human affairs. Even trips between neighboring stars, averaging several decades or more, fall within the scope of long-range planning, but as Michael Hart estimates in the next essay, crossing the Galaxy may involve at least a couple of million years. When we reach such a scale we enter the domain of geologic time. At this scale we can resolve events in the fossil record, and make available some of the data it contains, on more than 600 million years of change in animal life on earth through evolutionary and ecological processes. These data may serve as clues to the changes that might occur while an earthly species lineage spreads through the Galaxy.

Evolutionary Rates and Rhythms Suggested by the Fossil Record

The features of evolutionary rates that are of major interest include the length of time required for novel taxa, such as new species, to originate; the frequency of such originations; and the duration of the taxa, obviously related to their extinction rates. The fossil record of animals is longest and by far richest in number and variety for invertebrates of shallow sea habitats. All living phyla have representatives in the sea and as far as we can tell all phyla originated there. For these reasons both the quality and quantity of data on evolutionary rates are best for marine organisms, and so I shall draw on this information. I will then examine the question of how these data may apply to humans.

Species Lineages

The average duration of marine invertebrate species is on the order of 5 million years; the durations vary among taxa and with environmental setting. Bivalve Mollusca, a group with a relatively slow evolutionary tempo, seems to have average durations of around 7.5 million years per species, although in the late Cretaceous of the Western Interior, bivalve species tended to last around a million years on average. Ammonite species may have averaged as little as 1 or 2 million years in duration, although some long-ranging taxa that may represent single species last for 20 million years and more. The average duration of species may not represent an important modal point at all. Incidentally, vertebrates tend to have shorter species durations than most invertebrates; mammal species may have average durations of between 1 and 2 million years.

In contrast to the millions of years over which they commonly endure, the time required for the origin of a new species is often so short that we cannot resolve it in the fossil record, whether the species is an ammonite or a bivalve. Our understanding of the scales of resolution of events represented in the record is still quite incomplete, but it seems unlikely that we can regularly resolve processes that occur in a few thousands to a few tens of thousands of years. In theory, the ordinary processes of microevolution as now understood are adequate to produce a new morphospecies (a species defined on morphological difference as recognized in the fossil record) in thousands or perhaps only hundreds of years in extreme cases. Perhaps 10,000 years is a reasonable guess for an average fastest rate of speciation among all invertebrate lineages. A morphospecies that originated at such a rate would usually simply appear abruptly in the fossil record, without morphological harbingers. If a species then lasted for several million years before extinction, it would display the now-famous punctuational pattern: sudden appearance, little morphological change (or at least no strong directional trend leading toward a descendant species), and disappearance without obvious intermediates leading to later forms.

One of the important conditions for rapid morphological change is that the evolving population be small, so that a novelty may rapidly become the norm; small is sometimes taken as less than 100 breeding individuals. Very large populations, however, can usually evolve only slowly. Twin populations of a large number of shallow water invertebrates species, one on either side of the Isthmus of Panama, were isolated at least 3 million years ago when the Isthmus rose to cut communication between the Atlantic and Pacific oceans. The present descendants of these twin isolates can be distinguished from one another only with difficulty, although they are distinct species. Thus, millions of years may be required to produce morphospecies when the isolated populations are large and in similar environments.

For species that begin as small isolated populations with the potential for morphospeciation within, say, 10,000 years or so, the number of species that may be evolved is much greater than any observed number; a single ancestral species could produce more descendant species in a million years than have ever lived. Observed rates of the frequency of production of new species, just reviewed, tend to average close to replacement requirements; groups within which species last about 5 million years usually have observed speciation rates near 0.2 species/million years. It is abundantly clear that realized speciation rates fall far short of the potential rates.

Adaptive Space

A common way to view the adaptive opportunities in Earth's environment is to consider the environment as a mosaic of conditions, with each mosaic tile, or *tessera*, representing a region of relatively uniform conditions, separated from neighboring tesserae by boundaries where conditions change more rapidly and that therefore form adaptive barriers of various strengths. As each tessera includes a fraction of the world's resources, the number of tesserae is finite. Assuming that each tessera can be occupied by only a single species, it is possible to consider whether Earth's environment is or can be "full" of species—whether all tesserae may be occupied at the same time.

For any given environmental state, an equilibrium number of species (species diversity) can be established by an interplay of speciation and extinction rates. The nature of such an equilibrium will vary with the ecological assumptions of the model; according to MacArthur and Rosenzweig, competition is an important component of extinction, its per capita extinction effects increasing as diversity rises. An equilibrium diversity level is associated with a "full" biosphere then because any additional species will be extirpated by competition. However, there is little empirical evidence that competition plays such a role. Indeed, reviewed data on the invasion of ecosystems by immigrant species in more than 800 cases indicated that extinctions rarely ensued (less than 10 percent of the time) and that they could hardly ever be shown to involve classical competition. It appears that ecosystems ordinarily have room for additional species.

If one assumes that per capita extinctions are constant with respect to diversity, rather than increasing with diversity, then the environment is not full at equilibrium so long as there is some extinction. There could be ample room for added species. Timothy Walker has worked out the expression for diversity under this circumstance which is

$$\delta N/\delta t = [a(1 - N/N_{max}) - b]N,$$

where N is the number of species present, a the speciation rate, b the extinction rate, and N_{max} the number of tesserae in the environment and therefore the maximum number of species that the environment can contain; δN represents the change in N over a short time span, δt. At equilibrium,

$$\delta N/\delta t = 0 \text{ and } N = (1 - b/a)N_{max}.$$

By substitution of values of a and b determined from the fossil record, it appears that for most animal groups the environment is far from full; perhaps only about 70 percent of the tesserae are occupied on average. For a perspective on this figure, consider that the entire invertebrate fauna of the present oceans numbers about 350,000 species, so that there should be more than 150,000 empty tesserae to form opportunities for additional speciations.

This tentative finding, if corroborated, suggests why there is such a great disparity between the theoretical potential speciation rates and the observed rates—the adaptive barriers between tesserae are difficult enough to broach that speciation rates are held low, and extinctions keep the environment rather empty even at equilibrium diversity. It also suggests why species may appear so rapidly: They do not have to insinuate themselves into fully packed communities but may take over an empty tessera upon broaching the boundary.

Higher Taxa

Although species commonly appear abruptly, we might expect that the origin of higher taxa—phyla, classes, and orders—could be tracked in the fossil record, at least in some gross way, along a succession of species as they diverge from their ancestral group. In fact, this is not common. There is not a single invertebrate phylum or class that can be definitely traced through intermediate forms from an ancestral group. In some cases we have strong suspicions as to what the ancestral group was, but the higher taxa appear just as abruptly as do species, with their characteristic features—unique body plans or subplans—already evolved.

About thirty living phyla are recognized by most authorities. Some of these phyla are not even known as fossils, but these are composed chiefly of minute, soft-bodied forms that could be fossilized only under exceptional circumstances. Flatworms, for example, are believed to have given rise to higher invertebrate phyla and, if so, originated before 600 million years ago, but they have left no known fossils older than about 20 million years. But phyla with easily preservable skeletons have left us good fossil records. These groups begin to appear about 600 million years ago and all but one are known from the Cambrian Period. That one, the Bryozoa, has a skeletal ground plan that is not an obligatory accompaniment of the body plan of the organism; it is possible that the Bryozoa were present early in the Cambrian but did not produce mineralized skeletons until later.

Other phyla have body plans that require a mineralized, easily preservable skeleton as a coadapted feature of their form-function complexes. One such phylum for which strong evidence now exists is the Brachiopoda. Cloud pointed out long ago that unskeletonized brachiopods were unlikely organisms, so that the advent of brachiopod skeletons should correspond closely to the origin of the phylum. This suggestion has been strengthened by recent work on brachiopod lophophore function. The brachiopod skeleton is so closely coadapted with the form-function complex of the lophophore that the brachiopod body plan implies that the anatomical and skeletal ground plans were evolved in concert. Brachiopods are particularly interesting because in one of the best geological sections across the Precambrian-Cambrian boundary in Siberia, brachiopods literally lie upon the latest Precambrian rocks as part of the basal deposits of Early Cambrian age. There cannot be a large hiatus at this boundary, according to the Soviet stratigraphers. Four very distinctive orders of Brachiopods appear within the earliest Cambrian stage, but no Late Precambrian brachiopods are known from anywhere in the world. Thus, we can infer that the brachiopod body plan was developed from a non-brachiopod ancestor (probably a phoronidlike form) and diversified into major body subplans rapidly. My own prejudice is that phylum-level body plans may be evolved within hundreds of thousands rather than millions of years, since we have not been able to resolve the origin of well-skeletonized classes in the fossil record.

It is consistent with the fossil record, though not required by it, that all living higher invertebrate phyla, plus the chordates, evolved during a relatively short time span near the Precambrian-Cambrian boundary. In any case we have a history of the very rapid development of phyla going on for several million years and of few if any more being produced subsequently. There are quite a number of extinct phyla known—nearly twenty by conservative estimate—most of which first appear in Cambrian rocks. Some are well skeletonized and thus give us a good idea of when they became extinct. Some do not last through the Cambrian Period. Most, however, are soft bodied and are found only in exceptional fossil localities such as the Burgess shale (Middle Cambrian), British Columbia, or the Mazon Creek beds (Pennsylvanian), Illinois. These extinct groups do not appear to be ancestral to or intermediate between living phyla. These phyla presumably evolved during the same late Precambrian-Cambrian radiation that produced so many living phyla.

The record suggests that most phyla appeared abruptly during the most extensive radiation of evolutionary novelty the world has seen. Phyla may endure indefinitely (600 million years or so thus far), but there is no evidence that any has undergone a gradual succession of change that led to a descendant phylum. Many have become extinct, though we cannot say just when this occurs for most of them. Classes have a rather similar pattern, though the time of maximum appearance is shifted into the Ordovician and they continue to appear through the Paleozoic. For orders, the time of maximum origin is shifted even later and they continue to originate in post-Paleozoic times. For all these taxa the usual (though not exclusive) pattern is for the group to appear abruptly with its characteristic features assembled and often diversified into several subgroups.

These higher taxa thus have a punctuational pattern much like species in that their origins are quite abrupt compared with their durations. It is plausible to refer this pattern again to the structure of environmental occupation, with adaptive barriers preventing the exploitation of much of adaptive space; Bambach suggests that some regions of adaptive space may remain open for millions to hundreds of millions of years before being filled. Once the barrier is finally broached, however, radiation within the newly invaded adaptive zone may occur rapidly insofar as the invention of novel body plans and subplans is concerned. This pattern was described in 1944 by Simpson, chiefly from vertebrate evidence. The invention of whole new body plans, so common near the lower Cambrian boundary,* occurred in relatively wide-open adaptive spaces. There was a major diversification following the Permian-Triassic extinctions, which was nearly as great as the early Paleozoic ones at the level of species, genera, and families, so far as we can tell. However, it occurred in adaptive space containing scattered inhabitants and produced no phyla or classes.

Human Lineages in Galactic Space

The human species and its close allies have not exhibited an unusual evolutionary pattern. Species of *Australopithecus* and *Homo* seem to appear abruptly and most change relatively little so far as we can tell, and after perhaps a million years or so (more on average) they disappear. Hominids actually display slow evolutionary activity for mammals, but not extraordinarily so. Cultural traits aside, there is no reason to expect *Homo sapiens* to have a special evolutionary pattern. What then can we expect of our species when it is fragmented into many isolated populations, most of which are introduced into extremely novel environments while colonizing the Galaxy over a span of 2 million years? We will assume that much or all of the Galaxy is empty of close competitors.

The last thing we would expect is no changes. The colonization events, which would presumably involve small isolated populations

---

*It has been estimated that one in forty species originating at that time was at least a new class.

from the standpoint of evolution, are practically tailor-made experiments in inducing evolutionary novelty. Since they would number in the millions, there is every expectation that some of the "experiments" would succeed in producing distinctive new morphospecies, the origin of which should involve time scales of only thousands of years, even considering that humans have relatively long generations.

Speciation would require an evolutionary breakthrough into the adaptive conditions represented on a colonized planet or other body—a new unoccupied tessera or adaptive zone if ever there was one. There is no reason to expect less change than is usual between sister morphospecies that arise during "punctuational" events, and there is in fact some reason to expect more than average changes, at least in some cases. The assumed emptiness of space with regard to inhabitants of our adaptive zone evokes a comparison with the radiation of early Cambrian time. Then many of the new species did more than enter a new niche; they entered whole new adaptive zones that happened to be largely empty and underwent extensive morphological repatterning to create novel body plans. Some of those were so distinctive that we recognize them as phyla, classes, and orders, but they originated so rapidly that we cannot yet resolve their origin in the geologic record. There is thus probably plenty of time available within a 2 million-year period of expansion to permit the development of morphological types so unique as to merit assignment to new higher taxa.

We do not really understand the pressures and mechanisms required for the successful rapid shifts in morphology that are seen in the fossil record. There are a number of hypotheses, which need not be reviewed here. For our purposes, the morphological shifts may be regarded as spontaneous at frequencies sufficient to predict their occasional occurrence during a galactic colonization of distinctive and relatively empty adaptive space. Of more relevance than mechanism is the probable ability of humans to significantly influence their environments and their genomes so that morphospeciation will no longer be a spontaneous "natural" event. However, as Finney and Jones have pointed out, it might be doubted that every one of the hundreds of thousands of colonizing populations will be able or will choose to suppress evolutionary change; some might try to promote it; it would seem to be a local option. Incidentally, the extent of evolution that might be promoted in experimental organisms, especially if chosen for their evolutionary potential, would be nearly limitless.

It is sometimes suggested that evolution has run its course in our species. Assuming that galactic colonization occurs, experience with

274   JAMES W. VALENTINE

600 million years of animal life on earth suggests differently. Experience does suggest that many colonizing populations will
probably not change too much biologically, although to human
perceptions a minor evolutionary fluctuation might be all too obvious. But a certain percentage of the colonizing populations should
undergo significant changes that result in the sorts of morphological
differences that characterize different species of hominids. Some
human isolates may well undergo such extensive repatterning of their
morphology and physiology as to establish entreé into a new adaptive
zone. Furthermore, although conservatively we would expect extensive morphological shifts to be relatively rare, this does not mean
that such novel human descendant taxa will necessarily be rare in the
Galaxy. Any predilection or talent for colonization that an unusual
descendant taxon might have could lead to extensive spreading from
its founding colony and to its widespread occupation of galactic
space. By analogy with the established pliability of earthly genomes
early in the history of novel taxa, it would not be surprising if any
unusual human descendant lineage radiated farther into many
varieties within the timetable of galactic filling. Not that we should
expect the Galaxy ever to be truly full of hominids; extinction rates
should maintain a pool of unoccupied but habitable worlds, so that
even at an equilibrium occupation density, opportunity for colonization and further evolutionary novelty would never be foreclosed to
our descendants.

In conclusion, I fearlessly predict that within 2 million years the
descendents of *Homo sapiens*, scattered across the Galaxy, will
exhibit a diversity of form and adaptation that would astound us
residents of today's earth and that even then our evolutionary potential will hardly have been scratched.

References

Bambach, R. K. In press. Classes and Adaptive Variety: The Ecology of
    Diversification in Marine Faunas through the Phanerozoic. In
    J. W. Valentine, ed., *Phanerozoic Diversity Patterns: Profiles in
    Macroevolution*. Princeton University Press, Princeton.
Behrensmeyer, A. K., and D. Schindel. 1983. Resolving Time in
    Paleobiology. *Paleobiology* 9: 1-8.
Cloud, P. E. 1949. Some Problems and Patterns of Evolution Ex-
    emplified by Fossil Invertebrates. *Evolution* 2: 322-350.
Croizat, L., G. Nelson, and D. E. Rosen. 1974. Centers of Origin and
    Related Concepts. *Syste. Zool.* 23: 265-267.

Ekman, S. 1953. *Zoogeography of the Sea.* Sidgewick and Jackson, London.

Finney, B. R., and E. M. Jones. 1983. From Africa to the Stars: Evolution of the Exploring Animal. In J. D. Burke and A. S. Whitt, eds., *Space Manufacturing 1983, Advances in the Astronautical Sciences* 55: 85-103. Univelt, San Diego.

Kauffman, E. G. 1978. Evolutionary Rates and Patterns Among Cretaceous Bivalvia. *Phil. Trans. Roy. Soc. London B.* 284: 277-304.

Keller, B. M., Yu. A. Rezanov, V. V. Missarzhevsky, L. N. Repina, Yu. Ia. Shabanov, and C. E. Egorova. 1973. *International Excursion on the Problems of the Cambrian-Precambrian Boundary; Guidebook for the Excursion of the Aldan and Lena Rivers.* Akad. Nauk, Moscow—Irkutsk, USSR.

Kennedy, W. J. 1977. Ammonite Evolution. In A. Hallam, ed., *Patterns of Evolution as Illustrated by the Fossil Record.* Elsevier, Amsterdam.

Kennedy, W. J., and W. A. Cobban. 1976. Aspects of Ammonite Biology, Biostratigraphy and Biogeography. *Spec. Papers in Palaeontology* 17: 1-94.

LaBarbera, M. 1981. Water Flow Patterns in and Around Three Species of Articulate Brachiopods. *J. Exper. Mar. Biol. Ecol.* 55: 185-206.

Lande, R. 1980. Genetic Variation and Phenotypic Evolution During Allopatric Speciation. *Am. Nat.* 116: 463-479.

MacArthur, R. 1969. Patterns of Communities in the Tropics. *Biol. Jour. Linn. Soc.* 1: 19-30.

Pelman, Yu. L. 1979. Ancient Brachiopod Complexes (Class Inarticulata). In Zhuraleva, I. T., and N. P. Meshkoia, eds., *Biostratigraphy and Paleontology of the Lower Cambrian.* Translated by the Institute of Geology and Geophysics, Soviet Academy of Science, Siberian Division, Nauka, USSR. Pp. 34-39.

Rosenzweig, M. L. 1975. On Continental Steady States of Species Diversity. In M. L. Cody, and J. M. Diamond, eds., *Ecology and Evolution of Communities.* Belknap Press/Harvard University Press, Cambridge. Pp. 121-140.

Simberloff, D. S. 1981. Community Effects of Introduced Species. In M. H. Nitecki, ed., *Biotic Crises in Ecological and Evolutionary Time.* Academic Press, New York. Pp. 53-81.

Simpson, G. G. 1944. *Tempo and Mode in Evolution.* Columbia University Press, New York.

Speith, P. T. 1979. Environmental heterogeneity: A Problem of Contradictory Selection Pressures, Gene Flow and Local Polymorphism. *Am. Nat.* 113: 247-260.

Stanley, S. M. 1977. Trends, Rates, and Patterns of Evolution in the Bivalvia. In A. Hallam, ed., *Patterns of Evolution as Illustrated by the Fossil Record.* Elsevier, Amsterdam. Pp. 209-250.

Stanley, S. M. 1979. *Macroevolution, Pattern and Process.* W. H. Freeman, San Francisco.

Stanley, S. M. 1981. *The New Evolutionary Timetable.* Basic Books, New York.

Valentine, J. W. 1981. The Lophophorate Condition. In T. W. Broadhead, ed., *Lophophorates.* Sci. Studies in Geol: 5. Department of Geology, University of Tennessee, Pp. 190-204.

Valentine, J. W. In press. Biotic Diversity and Clade Diversity. In J. W. Valentine, ed., *Phanerozoic Diversity Patterns: Profiles in Macroevolution.* Princeton University Press, Princeton.

Valentine, J. W., and D. H. Erwin. 1983. Patterns of Diversification of Higher Taxa: A Test of Macroevolutionary Paradigms. In J. Chaline, ed., *Modalités, Rythmes et Mécanisms de L'Evolution Biologique.* Coll. Int. C.N.R.S. 330: 219-223.

"Methuselah lived to be 969 years old. . . . You boys and girls will see more in the next fifty years than Methuselah saw in his whole lifetime."

—*Mark Twain*

"All the star-borne colonies of the future will be independent, whether they wish it or not."

—*Arthur C. Clarke*

*18

Michael H. Hart

INTERSTELLAR MIGRATION, THE BIOLOGICAL REVOLUTION, AND THE FUTURE OF THE GALAXY

This essay has two main themes. One theme concerns the time scales involved in interstellar travel and migration and a comparison of those with other relevant time scales. (Estimates of several of these time scales are given in table 18.1.) The other theme concerns the interaction between the human expansion into space—the space revolution, one might say—and the even more important biological revolution, which also lies before us.

Time Scales for Interstellar Travel and Colonization

The road to the stars is wide and free; unfortunately, it is also distressingly long. Even by the most optimistic estimates, an interstellar voyage will take years to complete. Although technological advances have reduced the travel time between widely separated

TABLE 18.1   Some Time Scales of Interest

Event	Time
$T_{population} \times 2$	30 years
$T_{star\text{-}star}$	50 years
$<L>_{now}$	70 years
$T_{g \text{ minor genetic engineering}}$	100 years
$T_{cultural\ change}$	500 years
$<L>_{future}$	3,000+ years (?)
$T_{major\ genetic\ engineering}$	10,000 years (??)
$T_{natural\ evolution}$	1 Myr
$T_{galactic\ colonization}$	2 Myr
$T_{Galaxy\text{-}Galaxy}$	10 Myr
Age of Milky Way Galaxy	10,000 Myr
Collapse of universe (?)	60,000 Myr

points on Earth to mere days, often just hours, there is no reasonable hope that future technology will ever succeed in reducing interstellar travel times to months.

This impossibility is a direct, inescapable consequence of Einstein's theory of special relativity. Unless that theory is overthrown—and very few scientific theories are supported by such an overwhelming mass of evidence—no material object will ever be able to move faster than c, the speed of light.

In science fiction movies such as *Star Trek*, spaceships merrily whiz about the Galaxy at speeds much greater than c. In the real world it is only with the greatest difficulty and enormous expenditures of energy that we will be able to even approach that figure. The arguments of Purcell have convinced most scientists that travel speed of the order of 0.99 c are likely to remain infeasible, owing to the overwhelming amounts of energy that would be required. However, straightforward calculations show that once the energy of nuclear fusion is available, the fuel and energy requirements for travel at 0.1 c are not prohibitive. Because, as Dyson has pointed out, there are several propulsion methods that might be used to attain that speed, and because many scientists (such as Sagan, Jones, and Drake) have suggested travel speeds at least that high, I will in what follows employ 0.1 c as an estimate of feasible interstellar speeds.

Using that estimate and remembering that a typical distance from a single star to its nearest neighbor is about four light-years, we find that the time scale, $T_{star\text{-}star}$, for a short interstellar trip is about fifty years.

It will, though, take much longer than that to cross and colonize the Galaxy. In the first place, the Milky Way Galaxy (MWG, hereafter) is 100,000 light-years in diameter; so even if no stops were made en route, the trip would take 1 million years. Almost certainly, though, the colonization of the MWG will occur in stages, with pauses between each of the many trips.

It seems a reasonable guess that at first we will explore the solar neighborhood, sending expeditions to (and colonizing) the nearby stellar systems, and only later make longer trips. Eventually, our colonies will become large enough and well enough developed to send out expeditions of their own, exploring and settling uninhabited nearby star systems. We can therefore foresee a roughly spherical "frontier of colonization," expanding outward from Earth with a speed of $v_{colonization}$.

If there were no pauses between the founding of a colony and its sending out a new expedition, then we would have

$$v_{colonization} = v_{travel} = c/10.$$

But there will be such pauses. If we estimate that the typical pause will be roughly as long as the typical trip, then

$$v_{colonization} = 1/2\, v_{travel} = c/20.$$

At that rate, it will take about 2 million years to colonize the Galaxy. As there are a few hundred billion stars in the MWG, we might expect there to be at least that many separate civilizations in the Galaxy when the process of colonization is completed (perhaps far more, if the speculations of Jones and Finney about the settlement of interstellar comets are correct).

Now 2 million years is a long time by human standards; but compared with the age of the MWG (10 billion years) it is a very short interval indeed. If an earlier spacefaring civilization had ever arisen in our Galaxy—even as little as 20 million years (a mere one-fifth of 1 percent of the age of the Galaxy) before us—it would have completely colonized the MWG, including the Solar System and Earth, before the human race ever arrived on the scene. Quite obviously, this did not occur, from which we might reasonably infer that *we* are the first colonizing civilization in our Galaxy and for the moment probably the only species with an advanced technology. If this is so, it will be our descendants who are likely to colonize and populate the entire Galaxy.

Some scientists have come up with higher estimates for $T_{galactic\ colonization}$. But for the purposes of this discussion, it doesn't matter much if our estimate is a bit off. Even if $T_{galactic\ colonization}$ is 10 million years—or even 50 million years—it is still very much less than the age of the MWG and very much greater than any of the evolutionary time scales listed in table 18.1.

Will There Be Any Interstellar Empires or Federations?

Science fiction stories such as *Star Wars* often portray a Galaxy containing one or more vast interstellar empires. This might make for exciting fiction, but the real world of the future will not include any such empires. As pointed out in 1958 by Arthur Clarke, the long travel times involved in interstellar travel will prevent any would-be emperor from exercising sway over any significant portion of the Galaxy.

Consider, for example, the difficulties an empire would face in trying to control a region a mere 100 light-years in radius—less than one millionth part of the volume of our Galaxy. If an outlying colony revolted, it would be 100 years before the imperial capital even heard of the revolution, and at least 1,000 years before an expedition of reconquest or punishment could arrive! To hold an empire together under such constraints seems exceedingly difficult. As a practical matter, even a much closer colony, say one only 10 light-years (100 years travel time) away will be very hard to control politically. As Clarke puts it, "All the star-borne colonies of the future will be independent. . . . Their liberty will be inviolably protected by time as well as space."

By the same reasoning, any sort of interstellar rules or regulations will be almost impossible to enforce. Hence, (a) there cannot be any galactic federation, at least not one which exercises any control or authority; (b) nor could our Solar System (or any other one) be a wildlife preserve, as suggested by Ball, carefully protected by some interstellar empire or federation. The vast distances and travel times make it practically impossible to enforce restrictions against illegal settlements or poaching.

Indeed, even forms of interstellar cooperation that are purely voluntary and need no enforcement will be hard to establish and maintain. Any such conventions will necessarily have little flexibility because of the difficulty in getting changes agreed to. (Among other

things, this shows how useful it will be to have a reasonably unam-
biguous "galactic language"—that is, one to be used for interstellar
communication—agreed on before our expansion into space begins.)

Time Scales For Human Evolution

Human, *Homo sapiens*, evolved from earlier species, and (if we do
not destroy ourselves first) other species will arise from us. In the past
a new species arose only by "natural evolution," that is, natural
biological evolution on the Darwinian model. How long might it take
a new species to evolve out of *Homo sapiens* by that process?

We cannot, of course, give a precise answer to that question. Not all
new species take the same time to evolve. But to get a rough idea of
the time scale involved, we might consider the origin of *Homo
sapiens* itself. The immediate ancestor of our species was *Homo
erectus*, a species that was flourishing 700,000 years ago. From the
earliest specimens of *Homo erectus* to the first specimens of *Homo
sapiens* was an interval of a little more than a million years. So as a
very rough estimate, we might take $T_{\text{natural evolution}} = 1$ Myr.

However, although that might be a reasonable figure to use for past
changes, it is not the appropriate time scale to use for future evolutionary change. That is because we now stand on the brink of a new
era—the biological revolution, one aspect of which is genetic engineering.

Only a short while ago genetic engineering was the stuff of fantasy
and science fiction. Today it is a major field of scientific research, and
large-scale applications seem only a few decades away. The exact time
scale may be in doubt; but it seems clear that within a few centuries
we will be able to construct new genes (or combinations of genes) for
just about any reasonable human (or animal) trait or structure.

One result of this will be a rapid speeding up of human evolution.
Up to now evolution has been a slow process, constantly delayed by
the need to wait until some potentially useful new gene arose by
chance. With genetic engineering there will be no need to wait for
random mutation. If a new improved gene is desired, we can simply
build it! Under such circumstances evolution can proceed very
rapidly indeed.

But will genetic engineering actually be employed to alter human
beings? After all, there is already a lot of nervousness about genetic
research and resistance to its application on even nonhuman species. I
think the answer to that question is definitely yes. The history of the
past couple of centuries shows plainly that even if there is initially

some resistance, scientific breakthroughs are almost always quickly applied. Once an invention or process becomes technologically feasible, if there is any market for it at all it is soon put to use. This occurs even if no governments actively support the innovation; indeed even if many governments frown upon it. For better or worse, in the modern world technology seems to have a momentum of its own.

The first applications of genetic engineering to human beings are likely to be obviously benign. A typical early user might be a person who suffers from some hereditary disease or defect and who wishes his or her child to be spared the same handicap. Moderate governments will have a hard time preventing such applications of genetic engineering, if indeed they want to. After all, even apart from the humanitarian factors involved, people with hereditary defects are often a drain on the public treasury.

Once the safety of genetic engineering is established and its use becomes less unusual, it will be employed for more prosaic purposes, for example, to ensure that a daughter has blue eyes, to ensure that a son be six feet tall, or to provide a child with genes that will help give him or her an intelligence at least a bit above the current average.

Once the biological revolution is in full swing, minor changes such as these can be implemented quickly, although a given change might take about a century to become widespread. We might therefore estimate the time scale for minor genetic engineering changes as $T_{\text{ge minor}} = 100$ years. Major structural or biochemical changes will, of course, be much rarer, since social and governmental pressures are likely to oppose them. The time scale for such changes, which in effect produce a new species, is very hard to estimate, since the past provides little guidance. In principle, the use of genetic engineering could produce major changes in as little as a few hundred years. However, for purposes of discussion, I will adopt the estimate $T_{\text{ge major}} = 10,000$ years. That estimate is very uncertain, but in any event $T_{\text{ge major}}$ will probably be much less than $T_{\text{natural evolution}}$, ($\sim 1$ Myr) and therefore very much less than $T_{\text{galactic colonization}}$.

How Many Intelligent Species Will There Be in the Galaxy?

We can now see the first important interaction between the biological revolution and the space revolution. Since

$$T_{\text{ge major}} \ll T_{\text{galactic colonization}},$$

by the time our descendants succeed in colonizing the MWG they will have divided up into many different species (all derived from *Homo sapiens*, but by genetic engineering rather than by natural evolution).

Popular science fiction stories often portray the MWG populated by a wide variety of strange-looking, though intelligent races. For the reasons given in the second section of this essay, I believe this is probably an incorrect description of the present state of the Galaxy. It is, however, likely to be a fairly good description of what the Galaxy will be like a few million years from now. At that time there may well be hundreds of billions of different intelligent species in the Galaxy, many—perhaps most—of them bearing little physical resemblance to us.

We can also see how foolish it is to speculate what "their" culture will be like. There will be no single "they," but rather many different ones. The only reasonable prediction we can make about their cultures is that they will be extremely diverse.

The Coming Increase in Human Life Spans

The second and even more important aspect of the biological revolution will be an enormous increase in $<L>$, the mean human life span.

There is a currently popular theory that asserts that the principal effect of advances in medical science is to reduce death rates at lower ages (thereby raising the mean life span dramatically) but that the maximum human life span has been unaffected and probably cannot be raised significantly. I will call this the "rectification hypothesis," because it is usually illustrated by a graph like that in figure 18.1. That graph shows a "survival curve"; that is, the fraction of those persons born in a given year who are still surviving at the ages indicated by the horizontal axis. The shape of the survival curve changed markedly between 1850 and 1950, primarily because of reduced death rates in lower age groups. According to the theory, the process will continue, with the survival curve slowly approaching (though never reaching) the "ultimate" curve shown in figure 18.1, and thereby "rectifying the survival curve."

The rectification hypothesis, though rather orthodox today, is in my opinion totally incorrect, both empirically and in theory. It provided a reasonably good description of what occurred in the century before 1950, but we should not mistake a temporary phenomenon for an immutable law of nature. The theory does not correctly describe what has been happening in the Western industrialized countries during the last twenty or thirty years. Since about 1950

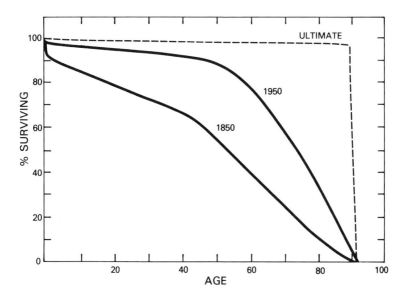

Fig. 18.1. Rectification Hypothesis. Between 1850 and 1950 the percentage of the population surviving to a given age increased for all ages, but most spectacularly at greater ages. However, with few exceptions the maximum age did not increase significantly. If this trend were to continue, the population curve might approach the dashed curve.

death rates in those countries have been falling most sharply in the higher age groups, and the most dramatic increases in future life expectancy have occurred at the higher ages. From table 18.2 we can see that between 1957 and 1978, for example, the largest percentage gains in future life expectancy have been for older people, not younger ones. We can also see that most of the gain in years of future life expectancy at age 10 is accounted for by the extra years expected after age 60 is reached, whereas only a small fraction of the gain at age 10 is due to the increased chance of reaching age 60.

In addition, the available evidence indicates that not merely the mean life span but also the maximum human life span has increased over the past two centuries. Back in 1800 a scientist investigating the matter might well have concluded that there was not a single well-authenticated instance of a human being reaching a 100th birthday. By 1950 there were quite a few such cases, the record being a man who lived to 113. Today there are many indisputable cases of persons living to 100, more than a dozen well-authenticated cases of people

TABLE 18.2   Changes in Future Life Expectancy*

Males

	1957	1978	Increase	Percentage increase
At age 10	59.4	61.5	2.1 years	3.54
At age 60	15.7	17.2	1.5 years	9.56

Females

	1957	1978	Increase	Percentage increase
At age 10	65.4	68.9	3.5 years	5.36
At age 60	19.2	22.3	3.1 years	16.15

*Average remaining lifetime, in years.
All figures are for white population of the United States.

reaching 110, and the record is up to 117. Taken together these figures indicate a substantial increase in L_{max}, with no obvious limit in sight.

But I do not wish to base my prediction that $<L>$ will greatly increase simply on recent trends in longevity. After all, that too could be a temporary phenomenon. Instead, let us try to consider the question of the maximum possible human life span from the standpoint of basic physical theory.

Aging, senility, and death involve a complicated set of physical processes, such as hardening of the arteries, formation of fat deposits in blood vessels, autoimmune reactions, formation of cross-links in protein, and many other processes, some of which are poorly understood (or even unknown) at present. But all of these varied processes, even the unknown ones, have one thing in common. They are all biochemical or biophysical processes, not mystical, incomprehensible changes. Therefore, each of them can in time be understood, and once we understand what causes a process, we can find a way to prevent or reverse it.

The notion that there is some cause of aging or death that is intrinsically incomprehensible is sheer mysticism, not science.

"But," you might ask, "what about the second law of thermodynamics? Isn't there something called entropy, which is somehow related to the disorder in a system and which never decreases? And doesn't that mean that if a change occurs in a system that increases the entropy, then that change must be irreversible?"

Indeed, there is a second law of thermodynamics, though it is often misunderstood by laypersons. It states that in a *closed* system—note the word *closed*—the entropy never decreases. Therefore, in a closed system, any process that increases the entropy (which in fact includes most processes of interest) cannot be completely reversed.

But according to the accepted laws of thermodynamics, if a system is not closed and if an external energy source is available, then any change occurring in that system is thermodynamically reversible. The human body, of course, is not a closed system. There are external energy sources available to us. (That in fact is what keeps us going.) Therefore, from the standpoint of thermodynamics all biochemical and biophysical processes occurring in the human body are reversible.

Many of you are perhaps still skeptical. You may have read that there are "biological alarm clocks" in the human body, perhaps in the individual cells, which are genetically programmed to go off at some age X (70? 90? 110?) and cause us to die. If that is so, you may think, there is no hope of us living much past age X.

There certainly are biological alarm clocks in the human body (for example, the ones that set off the changes occurring during puberty and at menopause), though it has not yet been clearly demonstrated that any such alarm clock sets off lethal changes. But let us assume for purposes of argument that there is such an alarm clock that, in the natural course of events, would kill us at age X. That still would not mean that L_{max} is limited to age X.

This is because any such biological alarm clock can only consist of some progressive biochemical or biophysical change occurring somewhere in the human body (such as, for example, the depletion of the original store of some chemical or the progressive increase in the amount of some chemical). Whatever the exact nature of the progressive change, the clock does not—cannot—involve any mystical process but merely some (as yet unknown) chemical or physical processes. Once we understand the relevant processes, we can reverse them, thereby resetting the "alarm clock" before it goes off. (Contrary to the common aphorism, it can be very nice to fool Mother Nature.)

From the foregoing, the following conclusions seem reasonable:

1. Over the course of the next few centuries, medical science is likely to discover essentially all the processes of human aging and how to prevent or reverse each of them.

2. When this occurs the mean human life span will increase enormously, and the maximum human life span even more.

3. Our descendants will not, however, be immortal. Individuals will still die eventually because of (a) accidents, (b) homicides (including suicides and war deaths), (c) failure to engage in proper health maintenance or aging prevention, (d) failure to obtain proper medical care when ill, or (e) the advent of a new disease, that kills some (or many) people before a cure is discovered.

These factors might restrict the future value of $<L>$ to a few thousand years. I will therefore adopt the estimate $<L>_{future} = 3,000$ yr, although considerably higher figures cannot be ruled out. However, there will no longer be any normal limit on L_{max} except perhaps the cosmological limit of A.D. 60 billion when the universe (if it is indeed closed) will be due to collapse back into a black hole.

The Effect of High Life Spans on the Frequency of Interstellar Voyages

We can foresee at least two ways in which the rise in human life spans will affect interstellar travel. In the first place, it will make such voyaging far easier.

At the present time a major obstacle to interstellar travel—indeed, the difficulty most often mentioned by those who doubt that it will ever be feasible—is that the time required for a voyage is greater than (or at least comparable to) an entire human lifetime. Several methods of overcoming this problem have been suggested, for example: (1) Put the voyagers in a deep freeze at the start of the trip and thaw them out shortly before arrival; (2) plan the voyage as one that will take many generations to complete. Both of these suggested solutions (as well as some others not listed here) may be feasible. Nevertheless, the problem may still discourage many would-be space voyagers and also those who might design, build, or finance the space vehicles.

If, however, a few centuries from now the typical human life span is thousands of years or more, the problem will become relatively unimportant. For persons who expect to live to 3,000 or beyond, a fifty-year voyage may seem like a minor interlude, and perhaps a very interesting one. They might well view it in roughly the same way as an 18-year old today views going away to college. In any event, a voyage to a nearby star system need not be viewed as either: (a) a permanent departure from one's home planet or (b) a trip begun without reasonable expectation of reaching the intended destination.

A second effect of long life span is that it will provide many people with a strong additional incentive to migrate. A world in which the average human life span is in the thousands will become over-populated very rapidly unless rigid restrictions on childbearing are adopted. Indeed, within a fairly short time a birthrate near zero will be necessary. What then will be done by those individuals who strongly desire children? Since they will not be permitted offspring if they stay in their home Solar System, they will be strongly motivated to emigrate to some empty, or at least sparsely populated, new system.

I am not suggesting that interstellar migration can ever solve society's problem of overpopulation. Only an effective birth control program can do that. All emigration can do is to solve the *personal* problem of some of those individuals who very badly want children, so badly that they will abandon their homes, friends, and planet to have offspring. However, once we expand into the Galaxy, there can be no central authority to enforce that unity, and the inevitable result will be the splitting up of the human race into many species.

The More Aggressive Races Will Populate
Most of the MWG

Suppose that no interstellar migration occurred and that all human beings remained within the Solar System. Then a central government (if one arose) might use its authority to place restrictions on the genetic engineering of human beings to maintain the unity of the human species. However, once we expand into the Galaxy, there can be no central authority to enforce that unity, and the inevitable result will be the splitting up of the human race into many species.

Again, within the confines of a single solar system a "world" government might use genetic engineering (or other biological tech-niques) to produce a less pugnacious, less aggressive population than *Homo sapiens*. In fact, in the MWG of the future, many individual civilizations may choose to do this.

I suspect, however, that it will be the more aggressive races that spread most widely through the Galaxy and that have the most offshoots. They will be the ones who explore most actively and migrate most frequently; they will be the ones whose members are most likely to enter strange and challenging environments; they will be the ones most willing to engage in interstellar warfare to displace another group from a desirable stellar system; and they will be the ones most willing to make large changes in their body structure or chemistry to adapt or to gain some competitive advantage.

As a result, most of the MWG will become populated by the more aggressive civilizations and species, and the long-range forecast is for continued aggression, war, and change.

How Much Interstellar Warfare Will There Be?

One should not infer from the foregoing that the average future inhabitant of the Galaxy will be faced with an interstellar war every few years. Quite the contrary, I believe that from the standpoint of most inhabitants of that future time, interstellar wars will be extremely rare, much rarer than warfare has been on Earth. This will be a consequence of the enormous distances between the stars and the large travel times between civilizations.

Here on Earth each nation has always been in close contact—through trade, tourism, migration, and the like— with its neighbors, and disputes have therefore been common. Furthermore, to settle such disputes by means of invasion has generally required a trip of no more than a few weeks. The result has been that a typical nation on Earth has experienced an average of one major war every fifty years.

In the galactic environment of the future, things will be very different. Each civilization will have relatively infrequent contact with its neighbors (other than by radio messages), and consequently relatively few disputes with them. Furthermore, invasions will involve much longer trips (fifty years or more), with correspondingly great expense and difficulty. Under such circumstances a typical civilization might be involved in a major war only once in 50,000 years, perhaps much longer.

There is a strange contrast here between the way the Galaxy will appear from within and how it might look to an outside observer. Because there are a few hundred billion stars in our Galaxy, if a typical circumstellar civilization is engaged in warfare for as little as five years out of a million, then at any given instant there will be a few million interstellar wars going on in our Galaxy. Outside observers looking at us might well say, "The Milky Way Galaxy is a jungle! A dog-eat-dog place where only the 'fittest' civilizations survive." Perhaps, on some cosmic scale they might be right.

Nevertheless, from the standpoint of average inhabitants of that "jungle," interstellar wars will be very rare. These inhabitants may well live long lives in a world that is basically peaceful and, by our standards, extremely prosperous.

References

Ball, John. 1973. The Zoo Hypothesis. *Icarus* 19: 347. Reprinted in D. Goldsmith, ed., *The Quest for Extraterrestrial Life: A Book of Readings.* University Science Books, Mill Valley, Calif. Pp. 241-242.

Clarke, Arthur C. 1958. *Profiles of the Future.* Harper and Row, New York. Chap. 10, especially pp. 118-119.

Drake, Frank. 1980. N is Neither Very Large nor Very Small. In M. D. Papagiannis, ed., *Strategies for the Search for Life in the Universe.* Reidel, Dordrecht, Holland. Pp. 27-34.

Dyson, Freeman. 1982. Interstellar Propulsion Systems. In M. H. Hart and B. Zuckerman, eds., *Extraterrestrials: Where are They?* Pergamon Press, New York. Pp. 41-45.

Hart, Michael H. 1975. An Explanation for the Absence of Extraterrestrials on Earth. *Quarterly Journal of the Royal Astronomical Society 16*: 128-135. Reprinted in D. Goldsmith, ed., *The Quest for Extraterrestrial Life.* Pp. 228-231.

McWhirter, Norris. 1983. *Guinness Book of World Records.* Chap. 1.

Purcell, Edward. 1961. *Radio Astronomy and Communication in Space.* U.S. Atomic Energy Commission Report, BNL-658. Reprinted in D. Goldsmith, ed., *The Quest for Extraterrestrial Life.* Pp. 188-196.

Sagan, Carl. 1963. Direct Contact among Galactic Civilizations by Relativistic Interstellar Spaceflight. *Planetary and Space Science* 11:485. Reprinted in D. Goldsmith, ed., *The Quest for Extraterrestrial Life.* Pp. 205-213.

SECTION V

Is Anybody Home?

INTRODUCTION

Our ultimate purpose in assembling the essays of this volume has been to speculate on our place in the universe. If we are alone, if the existence of intelligent, technological beings on our planet is an extraordinary accident of nature, then our future depends only on our ability to tame the universe and ourselves. However, if we are but the youngest of a multitude of technological species arisen in the billions of years of galactic history, then our future depends on how we might fit into such a community.

For the most part, the essays in the preceding sections have ignored the possible existence of extraterrestrial intelligence (ETI) and have either probed human experience for lessons that might apply to the coming human expansion into space or sketched technological developments that would make it possible. This was a conscious choice on our part.

Our goal has been to search for cultural (technology included) threads that might be followed into our future in space. The questions of whether ETIs exist and how we might detect them have been discussed at length in other conferences and books. Ultimately, these questions can only be answered experimentally: Either we will succeed in making contact or, after a thorough search, we will discover that we are truly alone. Perhaps there is a Galactic Club waiting for our membership application; or perhaps we are unique. We won't discover the truth by arguing about it.

Nonetheless, there are important questions involved. As in any human undertaking, a search for extraterrestrials means that we must make choices. What evidence are we looking for? What equipment do we need? How much would a reasonably thorough search cost? How long would it take? Can we estimate the chance of success? Those are practical questions. It is not surprising that they are hard to answer; informed guesses are the best we can do. In the fifteen years that followed the first serious SETI (Search for Extraterrestrial Intelligence) proposal by Cocconi and Morrison in 1959, the best guesses were that life was common and that intelligent life might be relatively easy to find—provided that some inherent flaw in technological youth did not condemn most or all of those who had come before us to early extinction. SETI proposals were designed around these assumptions. However, beginning with Michael Hart's paper of 1975, some have suggested that the conventional estimates are much too high. Hart asked the following question: If interstellar travel is practical it would seemingly lead to the rapid diffusion through the Galaxy of the first species to survive its technological youth. Why then, if extraterrestrials are so common, do we not see some sign of extraterrestrial colonists, or of their engineering projects, in the Solar System? (A version of this argument has been attributed to Enrico Fermi. We will return to Fermi's question at the conclusion of this introduction.) If Hart is right, it would have taken the survival of only one migratory species to fill the Galaxy; the chance of contacting someone would be far less dependent on the chances of surviving technological youth. One important factor in Hart's analysis is his estimate that an interstellar migration would fill the Galaxy in a time short compared with its age. If a migration takes a long time to complete, "they" might not have reached the Solar System yet.

It is a relatively simple matter to construct mathematical models of an interstellar expansion. The results of a particular model depend on choices of a few parameters, particularly the time it takes a new settlement to grow large enough so that voyages to more distant settlement sites could be undertaken. A journey between neighboring stars might take only a generation or so, but if successive jumps are separated by a wait of tens of thousands of years, crossing the Galaxy could take hundreds of millions of years. Estimates of waiting times depend less on technology than on social and biological factors.

The importance of social and biological questions to the description of an interstellar migration was a significant motivation for this book. In the following essay William Newman and Carl Sagan summarize the results of studies in population dynamics applied to an interstellar migration. They note that it is a statistical process, much like the treatment of motions of a gas. On a microscopic scale the molecules of a gas travel short, randomly directed paths between collisions with other molecules. The path of an individual molecule is chaotic. But when a gas is viewed on a larger scale, statistical (average) properties become important. We can describe mean forces such as pressure and average flow speeds that influence the motions of the gas on macroscopic scales. For an interstellar migration the voyages of ships from one star to another are like the motions of the gas molecules between collisions; "forces" tend to be things like population size and distance from the frontier. A very important "average" quantity is the rate of population growth.

Just how much averaging we are permitted in models of an interstellar expansion is a matter of some controversy. Newman and Sagan argue that the proper rate of population growth (a measure of the waiting time) is the average for the entire population. Others, including us, believe that growth rates on the frontier (which tend to be much higher) are more appropriate.

These are not simple matters. To return to the gas analogy for a moment, we note that the motion of a portion of a gas depends not so much on the properties of the molecules (they are all identical, at least in a statistical sense) as on local average properties like pressure gradients. In the case of an explosion the gas in the shock front is subject to strong pressure gradients; violent accelerations are the result. Back in the interior, pressure gradients are slight and the gas drifts outward at a steady rate. In other words, in a gas everything depends on local conditions. If we were to apply these observations directly to the case of an interstellar expansion, we could dismiss the use of behavior derived from averages over the whole human population and conclude that only growth rates on the frontier matter. However, human beings are not molecules in a gas. We are subject not only to local forces (accessibility to virgin territory, for instance) but also to social forces that can be transmitted over long distances. Because even the fastest interstellar migration would expand much more slowly than lightspeed, it is possible that ideas that might influence the pace of an expansion could spread from the saturated core of settled space to the frontier. Only a fool would dismiss the power of belief systems on human behavior.

In the opening paper of this volume we argued that exploration (and hence expansion) is a basic component of human nature, outweighing in many cases the effects of population pressure or concepts of territoriality. Newman and Sagan have a different view. They argue for the importance of population pressure and "the more aggressive motivations underlying colonization." In reality, all these factors and others undoubtedly have roles to play. One can point to specific examples in nature and in human experience that seem to support the dominance of any of the factors. Only the passage of time will sort out how the human expansion into space will proceed. We take an optimistic view: human are more explorers than aggressors. Opportunities are seized, and there is no opportunity more alluring than a frontier.

We began this short essay with a discussion of the implications of interstellar migrations for the search for extraterrestrial intelligence (SETI). In recent years opponents of NASA's SETI program used the arguments that we were alone to eliminate SETI funding from the NASA appropriation. Senator William Proxmire awarded SETI a "Golden Fleece." Fortunately, Carl Sagan and others were able to convince these opponents that these are questions that can be answered only by searching, not by talking. SETI funding was restored in 1983.

In the final essay, Jill Tarter outlines the awesome task of searching for extraterrestrials in the COSMIC HAYSTACK. Given limited resources, the searchers are faced with a multidimensional universe. Which stars are likely targets? Can we expect a steady signaling beacon, or must we hope to overhear chance conversations? Where in the vast electromagnetic spectrum might we expect signals? What kind of signals might they be? How would we distinguish artificial signals from the natural background? How might speculations about interstellar migrations influence the search strategies?

As Freeman Dyson has mentioned, a thorough SETI program would be indistinguishable from a thoughtful program of astronomical research. Perhaps we are alone, but we must make the effort. No matter how it comes out we will learn a great deal about the universe and our place in it. Ultimately, that is the goal of all human understanding.

Fermi's Question

Enrico Fermi was a frequent visitor to Los Alamos. There is a story about him that in some ways led to the conference and this book. It concerns a lunchtime conversation that took place in 1950 and it started when someone brought up the question of flying saucers.

Before the Second World War and the needs of the Manhattan Project brought some of the world's great minds to Los Alamos, the pine-covered mesas of the Pajarito Plateau bore little sign of man. There was a boarding school—the Los Alamos Ranch School—a rambling set of log buildings that surrounded a small pond. When the war came, and the military and the scientists came in the hope of ending it, the mesa tops were transformed. Labs and offices were built south of the pond and the main buildings of the Ranch School, particularly one called Fuller Lodge, became visitors' quarters and dining facilities. Although most of the World War II buildings have gone, and the business of the Laboratory is now conducted at sites scattered across mesas to the south of the original site, in 1950 Los Alamos was much as it had been during the war. Preparations were underway that summer for a series of nuclear tests to be conducted in the Pacific the following year.

One day a small group—Fermi, Emil Konopinski, Edward Teller, and Herbert York— started to walk over to Fuller Lodge for lunch. As Konopinski remembers it, "when I joined the party I found them discussing evidence about flying saucers. That reminded me of a cartoon I had recently seen in the New Yorker, explaining why public trash cans were disappearing from the streets of New York City. The New York papers were making a fuss about that. The cartoon showed what was evidently a flying saucer sitting in the background and, streaming toward it, 'little green men' (endowed with antennas) carrying the trash cans. In jest Fermi said that he thought that it was a very reasonable theory since it accounted two separate phenomena; the reports of flying saucers as well as the disappearance of the trash cans."

Teller remembers that although they all agreed that flying saucers were not real alien spacecraft, there was a brief discussion of possible modes of saucer propulsion. Was faster-than-light travel possible?

Fermi asked, "Edward, what do you think? How probable is it that within the next ten years we shall have clear evidence of a material object moving faster than light?"

"One in a million," Teller replied.

"This is much too low," said Fermi, "the probability is more like ten percent."

Teller mentions that this was a well-known figure for a Fermi miracle.

Konopinski recalls that as they made their way toward the Lodge, Fermi and Teller traded arguments back and forth, the numbers changing rapidly as they talked.

Once inside, the conversation moved on to down-to-Earth topics. Then, rather suddenly, "virtually apropos of nothing," Herb York tells us, "Fermi said, 'Don't you ever wonder where everybody is?' Somehow (perhaps because of the previous discussion) we all knew he meant extraterrestrials."

There was general laughter.

"It was his way of putting it that drew laughs from us," says Konopinski.

York continues, "He then followed up with a series of calculations on the probability of earth-like planets, the probability of life given an earth, the probability of humans given life, the likely rise and duration of high technology, and so on. He concluded on the basis of such calculations that we ought to have been visited long ago and many times over. As I recall, he went on to conclude that the reason we hadn't been visited might be that interstellar travel is impossible, or if possible, it is always judged not to be worth the effort, or technological civilization doesn't last long enough for it to happen." Although York confesses to being hazy on the latter part of the conversation, Teller recalls that the conversation may have included a statement that "the distance to the next location of living beings may be very great and that, indeed, as far as our galaxy is concerned, we are living somewhere in the sticks, far removed from the metropolitan area of the galactic center."

Fermi's question seems not to have gone beyond the confines of Fuller Lodge for a long time, although the issues implicit in the discussion were raised by others from time to time and in various guises. In 1966 Carl Sagan, having heard the story, mentioned without elaboration Fermi's question in a discussion of interstellar travel. And in 1976, after Michael Hart had brought these issues to wider attention with his provocative conclusion that we are alone in the galaxy, Sagan labeled the issue *Fermi's Question*.

References

Bracewell, Ronald N. 1974. *The Galactic Club: Intelligent Life in Outer Space*. W. H. Freeman, San Francisco.

Cocconi, Guissepe, and Phillip Morrison. 1959. Searching for Interstellar Communication. *Nature* 184: 844-846.

Hart, Michael H. 1975. An Explanation for the Absence of Extraterrestrials on Earth. *Q. J. Roy. Astron. Soc.* 16: 128-135. Reprinted in M. H. Hart and B. Zuckuman eds., *Extraterrestrials: Where Are They?* Pergamon Press, New York. Pp. 1-8.

Newman, W. I., and C. Sagan. 1981. Galactic Civilizations: Population Dynamics and Interstellar Diffusion. *Icarus* 46: 293-327.

Shklovski, I. S., and C. Sagan. 1966. *Intelligent Life in the Universe*. Holden-Day, San Francisco.

"Why undertake space flight at all? Why even consider a habitat away from planet Earth? The approach one takes in responding to such questions depends fundamentally on one's conviction about the destiny of humankind."

—*Joseph Allen*

*19

William I. Newman and Carl Sagan

NONLINEAR DIFFUSION AND POPULATION DYNAMICS

Mathematicians, biologists, and population demographers have developed nonlinear diffusion models as a method of quantifying the spread of animal (including human) populations. In this essay, we review these models and consider the manner in which such models can be extended to predict the spread of the human species by interstellar migration through the Galaxy. We then consider the implications for, the existence of, and the search for, extraterrestrial life.

During the last forty years, mathematical biologists have succeeded in developing methods for treating the behavior of an individual but representative member of a group, and from the behavior of the individual, deducing the most probable behavior of the group viewed as a whole. This work had its roots in the studies by such population geneticists and physicists as Fisher and Kolmogoroff and emerged from considerations of how advantageous genes would undergo spatial propagation. In studying the spread of various populations of mammals, population dynamicists have sought to isolate mechanisms that describe *in situ* growth, on the one hand, from its spatial propagation, on the other. To describe *in situ* growth most population experts employ a logistic law that embraces two elements.

1. At population densities far below the carrying capacity of the environment, population growth proceeds at an exponential, or Malthusian, pace. The growth rate is the difference between the birth and the death rates and describes the overall population gain in a year compared with the base population.

2. As the population grows it is limited by the carrying capacity of the environment and, assuming that the environment remains stable, reaches an equilibrium state. Should the population exceed this value, the environment will be incapable of supporting these added numbers and the population will decline. Thus, the equilibrium population can also be regarded as a saturation value.

The spatial propagation of mammals is referred to as dispersal, which is characterized by the animal heading off in a randomly selected direction over some characteristic distance before it mates, reproduces, and dies. Since inbreeding is in this way avoided, the frequency of genetic recombination of deleterious recessive mutations is reduced. Although it is arguable whether this is the reason for observed dispersal, recent evidence points toward population pressure as a significant causative mechanism. From a probabilistic point of view, each animal executes (in the absence of population growth) a random walk or Brownian motion. Among more intelligent animals, displacement from the point of birth is not entirely random but a consequence of diverse influences such as the need to reduce competition for food with others of its kind. Mathematical biologists have developed models that embrace both facets of animal behavior, models that have the form of either partial differential or partial difference equations.

The mathematical methods we have alluded to are designed to describe the *probability* of finding a member of this population at a given position at a given time. Although individuals exhibit marked variations in the speed of spatial propagation, the order in which spatial niches are established and inhabited, and so on, the probability associated with the spread of a population will appear characteristically smooth. In a sense, the probability associated with finding a member of this population at some site may be associated with the average or expected value of the population density. (A useful analogy emerges from the game of dice. The result of individual throws of a single die vary from one to six; the average value of a large number of throws is 3.5.) Although knowing only the probability of finding a member of the population to a given position obscures the behavior of individuals, it clarifies the behavior in the large of the entire population. The most important result to emerge from these population models is that the probability of finding a member of this population has a sharp spatial edge, indicating to a significant degree that the population is largely confined to some spatial region and that this edge or "population front" propagates at a uniform speed. (When Fisher first developed the population genetics model, he referred to this behavior as "the wave of advance of advantageous genes.")

In recent years mathematical biologists have increased the scope of this investigation from genetically oriented problems to the interaction of several species of animal populations with more complex logistic laws and dispersal mechanisms. What is significant to this discussion is that these population dynamics models possess the same underlying features, namely, a spatially confined wave front for the probability (of finding an individual member of a population), which advances spatially at a uniform speed. An apparently straightforward next step is to adopt similar methods for describing the behavior of representative members of a human population and to deduce the most probable behavior of the group considered as a whole.

It is not difficult to develop models for *in situ* growth and dispersal of human populations. The dispersal mechanism need not be random, as it is to a large degree in animal populations, but may be highly directed toward sparsely populated, desirable locations as a response to population pressures. (Early models of animal dispersal were strictly random, or Brownian; models that accommodate population pressure are characteristically nonrandom and in fact have been described in the literature as "directed motion" models. As before, the predicted spatial propagation of populations is wavelike with a sharp, confining front. Although the mathematical problems associated with such models are relatively well understood, more critical attention must be paid to the validity of applying such models to human populations.

The evolution of animal populations is clearly governed by what might be called a "Darwinian ethic"; in that view these population models are merely mathematical representations of the biological mechanisms that exist to ensure the survival of species. Humans (and intelligent extraterrestrials, if any) may behave in a much more complex fashion; their actions are sometimes governed by more than simple biological concerns. The question that emerges then is whether this sort of human behavior can be modeled as being, in the large, predictable. If human history must be viewed as a collection of near calamities, then the future of our species cannot be foreseen. We argue that human history supports the notion that such human behavior can be modeled and that human experience might be extrapolated into the distant future when humans will be capable of colonizing distant star systems. An essential part of the problem of deciding whether human history is largely predictable is our innate tendency to focus on isolated events. As an illustration, figure 19.1 shows an estimate of the population of Europe during the period of recurrent episodes of the bubonic plague. This figure clearly illustrates the conventional wisdom that the population of Europe

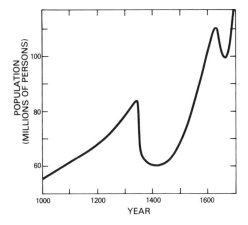

Fig. 19.1. Impact on population of recurrent plagues in Europe. Reproduced with permission from W. L. Langer "The Black Death" *Scientific American* 210 (1964): 117.

experienced a drop of one third during that calamitous period. (There is more recent evidence, however, garnered from manorial records of that period, indicating that the population drop may have been less severe.) In figure 19.2, we plot the population of Germany from 400 B.C. to the present. Here the influence of the bubonic plague appears to be a minor perturbation on the overall growth of Germany's population. When the global population growth is examined, the plague is hardly detectable. We conclude that when we move from

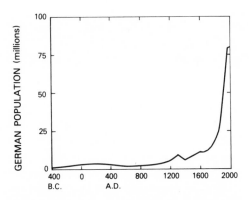

Fig. 19.2. Estimated population of Germany from 400 B.C. to the present. Reproduced with permission from C. McEvedy and R. Jones, *Atlas of World Population History* (Harmonsworth, England: Penguin Books, 1978), p. 69.

narrow focus to broad-scale and long-term perspective, history takes on the character of a sequence of events whose behavior in the large is smooth. (This observation is not unlike the difference discussed earlier between the behavior of individual members of a population and the overall behavior of the group.) Thus, we regard key events in history not as singular events but as developments that maintain a uniform rate of change over millennia. Indeed, we argue that in many cases this observation can be extended into the distant future; however, other hazards in the figure (e.g., a global nuclear or biological war before the establishment of self-sufficient human outposts on other worlds) could abruptly terminate our extrapolation.

A natural question that emerges in terms of population modeling is whether there are any examples drawn from human history of the spread of human populations having much the same character as the spread of animal populations, that is, a well-defined front of wavelike propagation. Figure 19.3 documents how the bubonic plague proceeded through Europe in a steady, wavelike fashion. Although the plague represents a disease resulting from the interaction of human and animal (i.e., flea and rat) populations, it is the routine motion of humans in the course of ordinary economic, military, and other activities that resulted in the spatial propagation of the disease. Figure 19.4 illustrates how the spread of agriculture proceeded across Europe from the Near East during the Neolithic period, again in a wavelike fashion. We have shown elsewhere that the colonization of

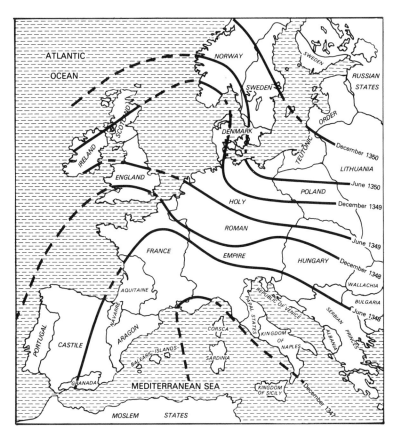

Fig. 19.3. Approximate chronology of the rapid sweep of the Black Death through Europe in the mid-fourteenth century. Reproduced with permission from W. L. Langer "The Black Death," *Scientific American,* 210 (1964): 116.

America, when proper account is taken of topographical considerations, and the like, proceeded overall in a wavelike fashion with a propagation speed of around 10 km/yr. Having argued that some large-scale aspects of human history might, in some sense, be predictable, we extend this notion into the future in attempting to describe the colonization of space and interstellar migration.

We now consider a time, perhaps several centuries hence, when humans have avoided nuclear and other means of self-destruction and have explored and colonized the habitable worlds of our own Solar System, and of other stars. Eventually, established colonies will muster their own interstellar exploration, and the human species will be slowly spreading through the Milky Way Galaxy.

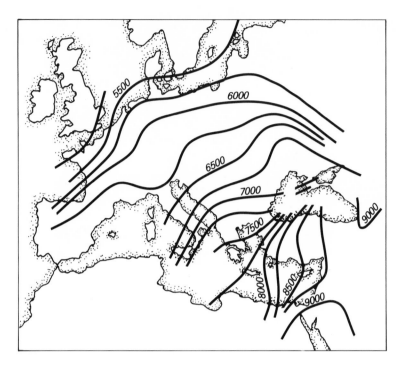

Fig. 19.4. The spread of agriculture across Europe from nuclear areas where agriculture was known more than 9,000 years ago. Reproduced with permission from L. L. Cavalli-Sforza "The Genetics of Human Populations," *Scientific American,* 231 (1974): 88-89.

Following the procedures established by population biologists (and in the context of interstellar colonization by Eric Jones and ourselves), we assume that a description of individual colonization efforts can be characterized by three principal elements.

1. Colonization ventures are most likely to be launched from established population centers. Before becoming self-sufficient and independent, we think it unlikely that a population center will sacrifice valuable resources (both human and material) to new colonization ventures.

2. Colonies will be launched from existing population centers to the nearest viable site. Contrary to the romantic notions shared by some that interstellar travel will be inexpensive and commonplace, we feel that space travel and the launching of

colonies will make very substantive resource demands on existing population centers, demands that will result in a parsimonious selection procedure for colonization sites, especially with respect to their distance.

3. New colonies will be established preferentially on unpopulated worlds, in part to enhance the information gained in the experience and in part to satisfy the desire to inhabit a virgin world.

Jones first explored this model using Monte Carlo techniques; our work was primarily analytical and considered a much broader array of parameters than did Jones. Fortunately, the results of our work can be derived approximately in a very simple fashion.

First, we wish to distinguish between the speed of an individual spacecraft, say, v_{ship}, and the speed of the colonization wave, v_{col}. In particular, the speed of the interstellar spacecraft can be expected to be much higher than that of the colonization front. To see this we note that

$$v_{col} \sim \frac{distance}{t_{travel} + t_{col}}$$

where the distance referred to is that to the nearest viable site and equals the ship's speed multiplied by the travel time t_{travel}. The time t_{col} represents the time it takes for a virgin world that has just received a group of colonists to grow in population to self-sufficiency so it is capable of launching a colonization venture to a new world. Since we expect that t_{col} is much greater than t_{travel} for interstellar spacecraft traveling at, say, 1 percent the speed of light (c), it follows that v_{col}/v_{ship} approximately equals t_{travel}/t_{col}. The colonization time scale is, crudely speaking, a measure of the number of e-foldings* that the population undergoes from its initial colony size, say, N_{col}, to the time when its population is self-sufficient (with population N_{suf}) and is capable of launching its own colonization venture. Thus, where γ is the population growth rate (in an exponential logistic law), $1/\gamma$ is the e-folding time and, approximately,

$$t_{col} = 1N\,(n_{suf}/N_{col}/\gamma$$

---

*The "e-folding" time is the time during which a population increases by a factor of $e = 2.7...$, the base of the natural logarithms. This is then the characteristic growth time of an exponentiating population.

Assuming $v_{ship} = 0.01$ c, we might expect a travel time of 1,000 years. (For such slow ship velocities, we are imagining the metabolism of the colonists reduced to minimal values during transit or a large vessel in which the colonists who arrive are the distant descendants of those who set out. Eventually $v_{ship} > 0.1$ c seems likely.) The population that a colony must possess before it can support a venture to another world is quite uncertain; fortunately, this has little effect on the overall result. Assuming that a population base approximately as large as the United States is needed to launch an O'Neill-sized venture with 10,000 colonists, we obtain $\ln (N_{suf}/N_{col}) \sim 10$, from which we deduce that

$$v_{col}/v_{ship} = 0.1 \ \gamma \ t_{travel} \sim 100 \ \gamma.$$

From this expression, it also follows that, as the time it takes to cross the Galaxy in a spacecraft traveling at 0.01 c is 10^6 yr, the time to colonize the Galaxy, with no other impediments, would be $1/(100 \ \gamma)$ times greater, or $10^4/\gamma$ y. The net human population growth rate (γ) has a strong influence on the time needed to colonize the Galaxy.

Currently, the human population is growing at a rate of 2 percent per year ($\gamma = 0.02$); most demographers, however, believe this is largely the result of a short-term imbalance between the global birthrate (which has not changed significantly over many generations) and the death rate (which has undergone a dramatic decline, especially in developing countries, primarily as the result of reductions in infant mortality and improved medical care). Almost certainly, the present circumstance will be followed by a resumption of the close balance between the birth and death rates, for which there is some evidence in technologically advanced countries ("the demographic transition"). Although the net population growth rate during most of human history was about 6×10^{-4}/yr and likely much lower in prehistoric times, a reliable prediction of future values of γ is impossible at present. (For example, if zero population growth were to become widespread, we would obtain $\gamma \sim 0$.) In the face of this uncertainty, it is impossible to establish firmly the time scale for the human colonization of the Galaxy. Should the growth rate return to the very low values (although for different reasons) characteristic of most of human history, this time period would be $\sim 10^8$ years or perhaps longer. If, on the other hand, prevailing global growth rates were to persist, the Galaxy could be fully colonized in 10^6 years.

What is especially significant about these time scales is their extraordinary length compared with the lifetime of human societies and with characteristic evolutionary (i.e., speciation) times. Over such time scales it is no longer clear that even gross predictions of population growth and the propagation of emigration wave fronts are possible. Indeed, we may bear little resemblance to our descendants 10^8 years hence; we can say even less about their aspirations, motivations, or intellectual attributes. One thing, however, is certain: To be able to talk about our descendants many generations from now, we must guarantee the survival of the human species on Earth during the next few centuries. During that critical period of time, small groups and possibly powerful individuals will have the capacity to destroy all human life on Earth. If the human species is to survive the invention of weapons of mass destruction, there must be a fundamental and universal change in human behavior. After that period of time, should colonies be firmly rooted elsewhere, we may be guaranteed our long-term survival. However, the requirement that we survive this critical period implies that our descendants and their society likely will be significantly different. One probable outcome of this transition in human behavior is a substantial quenching of the more aggressive motivations underlying colonization. Any substantive change in human behavior and aspirations may have a profound impact on humanity's migration to the stars.

Having explored some of the ideas underlying multiplicative diffusion models for population dynamics and their implications for interstellar migration and colonization, we now briefly consider possible implications of this model for the existence of extraterrestrial life. For many years considerations of the multiplicity of solar systems and the evolution of life have led scientists to speculate that life should be abundant in the Galaxy, a natural outcome of the belief that our Solar System and prebiotic chemistry are not atypical of what might be found elsewhere. Together with that understanding, some scientists hold that the emergence of intelligent life elsewhere would result in spacefaring societies that would rapidly colonize the Galaxy and that if extraterrestrial intelligence is abundant, we should expect to see evidence of extraterrestrial life in our own system.

The detailed calculations that we have performed elsewhere indicate, as mentioned earlier, that the time scale for colonizing the Galaxy is very long, so long that biological evolution (as well as shorter-term pressures for survival) would markedly alter the nature of that society. We, as well as other scientists, have advanced numerous explanations for why we have not so far seen any signs of extraterrestrial life. Our contention is, especially in view of the meager effort invested by American, Soviet, and other scientists to find evidence of extraterrestrial intelligent life, that it is premature to make an assessment of the abundance of extraterrestrial life. Absence of evidence is not evidence of absence. The search for life beyond Earth is an issue of profound importance. Many people have an emotional investment in the outcome of this question that touches on religious and political matters where predispositions have traditionally played important roles. But it is abundantly clear from the history of science that no convincing resolution of this issue is likely to come from protracted debates carried on with great passion and sparse data. We have an alternative denied to medieval scholastics: We are able to experiment. We can organize a scientifically rigorous, systematic search for extraterrestrial intelligence using the technology of modern radio astronomy. That is where the energies should be focused of those concerned with the great issue of the existence of other technical civilizations in the cosmos.

References

Jones, E. M. 1982. Estimates of Expansion Time Scales. In M. H. Hart and B. Zuckerman, eds., *Extraterrestrials: Where Are They?* Pergamon Press, New York.

Newman, W. I. 1980. Some Exact Solutions to a Nonlinear Diffusion Problem in Population Genetics and Combustion. *J. Theor. Biol.* 85: 325-334.

Newman, W. I. 1983. The Long-Time Behavior of the Solution to a Nonlinear Diffusion Problem in Population Genetics and Combustion. *J. Theor. Biol.* 104: 473-484.

Newman, W. I., and C. Sagan. 1981. Galactic Civilizations: Population Dynamics and Interstellar Diffusion. *Icarus* 46: 293-327.

Sagan, C. 1982. Extraterrestrial Intelligence: An International Petition. *Science* 218: 426.

Sagan, C., and W. I. Newman. 1983. The Solopsist Approach to Extraterrestrial Intelligence. *Q. J. Roy. Astron. Soc.* 24: 113-121.

"I have told this story of the greening of the galaxy as if it were our destiny to be nature's first attempt at an intelligent creature. If there are other intelligences already at large in the galaxy, the story will be different. The galaxy will become even richer in variety of life styles and cultures. We must be careful not to let our wave of expansion overwhelm and disrupt the ecologies of our neighbors. Before our expansion beyond the solar system begins, we must explore the galaxy thoroughly with our telescopes, and we must know enough about our neighbors to come to them as friends rather than as invaders. The universe is large enough to provide ample living space for all of us. But if, as seems equally probable, we are alone in our galaxy and have no intelligent neighbors, earth's life is still large enough in potentialities to fill every nook and cranny of the universe."

—*Freeman Dyson*

❋ 20

Jill Tarter

PLANNED OBSERVATIONAL STRATEGY FOR NASA's FIRST SYSTEMATIC SEARCH FOR EXTRATERRESTRIAL INTELLIGENCE (SETI): IMPLICATIONS OF THIS CONFERENCE

For more than a decade scientists associated with NASA have been trying to refine the strategy by which existing radio telescopes and state-of-the-art (SOA) digital signal processing hardware designed specifically for SETI could be employed to conduct a systematic search for evidence of extraterrestrial technology. During the decade there has been an explosive increase in the computational capacity afforded by a nominal SOA piece of hardware. No end is yet in sight for the Silicon Valley revolution; what we can do this year could be done better and cheaper next year. But SETI, as currently envisioned, is running out of time because of the rapidly increasing use of much of the electromagnetic spectrum for terrestrial communication. In particular, a search at centimeter and millimeter radio wavelengths may be precluded from the surface of the earth by the end of this century owing to the proliferation of transmitting satellites in geosynchronous orbit. The time to begin SETI is now.

This essay briefly describes the rationale for a search at radio frequencies and outlines the planned observational strategy. The intent of this strategy is to achieve an optimal exploration of the multidimensional search space inherent to the task (the so-called COSMIC HAYSTACK) consistent with a pragmatic limitation on funding and resources.

The essays in this volume raise the following questions: In the event that interstellar migrations are the norm for other civilizations, should the near-term observational strategy be modified in any way to increase the probability of a successful detection? Assuming the failure of this first systematic observational program (properly modified), what future strategy should be adopted? Finally what, if anything, can be concluded about the existence of extraterrestrial civilizations, about the inevitability of interstellar migrations, or about our ability to ever answer these questions?

Why is a paper on the strategy for a radio frequency SETI observational program appearing in a volume devoted to the problems and potentials afforded by the possibility of interstellar migration? For some people, in some places, it has become fashionable to state that the ancient question of the existence of extraterrestrial intelligence is intimately connected to the concept of interstellar migration; the existence of the latter implying the nonexistence of the former. Indeed, it is possible to construct a logical paradox embodied in the question "Where are they?" It is fitting that the modern revival of this query is attributed to Enrico Fermi during the time he spent at Los Alamos. History does not report whether Fermi took his logical exercise seriously, but others certainly have done so. Fermi's question ("Where are they?") has provoked a vigorous debate, often couched in detailed discussions of specific loopholes, but I prefer to review the structure of the logical argument in more general terms.

In its most succinct form, the Where are they? paradox may be rendered as

1. *IF* extraterrestrial civilizations have existed elsewhere and "elsewhen" in our galaxy,

2. *AND IF* interstellar travel/colonization/migration is inevitable for at least one of them,

3. *THEN* simple calculations indicate that an expanding wave of colonization will fill the galaxy on a time scale short compared to the lifetime of the galaxy,

4. *BUT* we do not "see" them here,

5. *THEREFORE* (1) is wrong; there has never been another technological civilization anywhere or "anywhen" in our galaxy except the Earth!

Ultimately, this may turn out to be the correct conclusion, but on the basis of data on hand at this time (including some of the discussions presented elsewhere in this volume), it could as well be that assumption 2 is wrong. Is space travel/colonization/migration really inevitable for any technological civilization? There have been numerous indications in the presentations of this meeting that many of the more obvious problems and pitfalls to colonists and colonization can be avoided. Joseph Birdsell argues that conscious selection to maximize the biological distances between partners can successfully avoid potentially disastrous inbreeding, even in a breeding population initially as small as sixteen. Genetic (geographical) diversity in the original membership of a multigeneration space colony is an additional safeguard, and so it would seem that very long interstellar space flights might be sanely conducted in this manner. Inbreeding concerns vanish entirely if genetic engineering eventually produces individuals with life spans sufficiently long that a colonization journey may be undertaken as merely one phase of an effectively immortal existence (Hart's essay). Interstellar comets or unintentional interstellar arks may provide a relatively economical means of making the long jumps between stars (Jones and Finney; Wachter) and thereby quell the cries of those who bemoan the exorbitant costs of such ventures. Indeed, some profit motive of such flights can be demonstrated, along with low-gravity resources to supply them (Criswell). It may even be possible that with sufficient preplanning and preadaptation, future descendants may handle the difficult interactions with alien species or with their own divergently evolved kin in a more graceful manner than has characterized such interactions on the face of this planet (Crosby). Last, any moralistic or legalistic concerns for the unborn whose parents undertake such a decisive and unidirectional voyage may be at least partially argued away by analogy with historical human behavior on the planet's surface (Regis). But these arguments, although reassuring, cannot ensure that interstellar migration will be inevitable.

There are still unknowns: physical, astrophysical, and psychological. Even beyond that, WE don't do everything that we could, and maybe THEY never do either, for some good and sufficient reason embodied in some form of universal morality. Other presentations elsewhere in this volume are as sobering as the discussions just mentioned are reassuring. Contrary to popular belief, not all ecological niches need be filled, even by an aggressively adaptive life form (Valentine). Indeed, the most aggressive forms may never possess the luxury of time needed to perfect the technology that would ensure that interstellar migration is a valid assumption. This latter possibility has been discussed elsewhere by Brin.

It is equally plausible, however, to say that assertion 3 (rapid expansion) is the incorrect fragment in the logical structure. The calculational models employed to date are still somewhat simplistic. The strength of the resulting temporal discrepancy in the logical construct results primarily from the unsaturated growth of a robust exponential function. This may indeed prove to be the correct model. However, papers at this meeting have reminded us that not all colonial ventures succeed (Lee; Schwartz), even when they entail only relatively modest extrapolations of the colonists' prior circumstances. It is the first colonies that are particularly vulnerable to catastrophic failure (Crosby) and indeed the mathematical models are the most vulnerable during these first finite steps. Finally, it is possible to assume that all colonies do succeed and still reach conclusions that do not contradict the apparent absence of local colonists even after galactic time scales. It is only necessary to assume that the correct parameterization for the model calculations is a more conservative one (Newman and Sagan). The models serve as a useful tool but should not be assumed to accurately portray the reality of galactic demographics.

Although at first glance the statement 4 (we don't see them) appears to be the single unassailable link in the logical chain, how well have we really looked? It may be fair to assert that we have not seen within our Solar System any evidence for the class of extraterrestrials (ETI) who compulsively engage in large-scale astroengineering projects. But David Brin's "principle of diversity" reminds us that this is not the only possible behavior for colonists; perhaps our local variety of colonist gets its kicks in some other fashion. Also, who is to say that our Solar System still possesses all the condensed raw material it started with? Would we really miss an asteroid or a comet or two? There have been several tentative identifications of asteroids hosting at least one companion object. Might this be the mining platform of a colonial enterprise? We cannot answer even this relatively well-posed and specific observational question; it seems presumptuous to assert that we do not "see" them when we have made only limited progress on a systematic exploration of our own Solar System. THEY may well be here (although I do not believe so) for all the looking we have done.

Item 5 (the uniqueness conclusion) is really what is wrong with the entire Where are they? logical construct. Popular publications and sporadic hit-or-miss searches of the past two decades notwithstanding, we have at the moment no data or basis on which to attempt to draw any cosmic conclusions. The question of the existence of extraterrestrials is properly an observational and not a theoretical one. SETI is our current attempt to try an experiment that might just answer one limited version of the grand question of life in the universe; that is, can we currently find evidence of the existence of extraterrestrial technology? In the following sections I outline the SETI strategy now envisioned and then try to indicate how the concept of interstellar colonization affects that proposed search strategy.

SETI: Observational Strategy

Current plans call for a search at microwave frequencies in one of two observational modes: the target search and an all-sky survey; in other words, a search at certain radio frequencies that either picks out directions of interest on the sky *a priori* (other stars like our Sun) or attempts to cover all directions on the sky, spending less time on any one direction.

Why radio waves? Why not gravitons, neutrinos, or other exotica? It is anticipated that any signal between civilizations (even between colonies and their place of origin) must travel across the large distances that separate stars in our galaxy. Therefore, the carrier for such a signal should be easy to generate and encode information on, should be economical, and should propagate through the interstellar medium without being deflected or scrambled. Any form of charged particle will be misdirected by the weak magnetic fields that thread throughout our galaxy. Any massive neutral particle will require extraordinary expenditures of energy to accelerate close to the speed of light and possess a nontrivial cross section for absorption along the way. The collection of nearly massless neutral particles (known and conjectured) has extremely small capture cross sections and thus could propagate over long distances without being deflected, but for this very reason are extremely difficult to generate, modulate, and receive. If the signals we seek are being transmitted in this fashion, then no search undertaken with our current technology will be successful. However, there is an easier way. We have found it very effective to use electromagnetic radiation (photons) for communication purposes on Earth and in space, and photons satisfy all the desirables for a signal carrier. We can attempt a search for photons with our current technology; but for what photons—that is, at which wavelength—should we search? We can currently detect electromagnetic radiation over 14 decades (14 factors of 10) of frequency, or wavelength, and this is enough. Frequencies much higher than gamma rays and much lower than meter-length radio waves cannot propagate very far through the interstellar medium. Among all these decades, preference for microwave frequencies arises not from the state of our own technology but rather from a census of the level of naturally occurring background noise at any position in the Galaxy owing to sources of astrophysical radiation. The microwave region of the electromagnetic spectrum is the quietest portion. If the civilization sending the signals wishes to minimize the energy required to transmit a bit of information, they will choose to use microwaves as they will have to "shout" the least loud to be heard above the cosmic noise at these frequencies.

Having specified "how" to look and listen (for long times at nearby stars like the Sun and for much shorter times at every direction on the sky) and "where" to look (in the microwave region), the search effort required should be well defined, and it should be possible to get on with the task. Unfortunately, things are not that simple! When searching for a signal there are many other parameters to be considered, some of them pertaining to the unknown nature of the transmitted signal itself. It is not enough to be looking at the right place on the right frequency; one must be looking at the right time and be prepared to detect signals of the right type with an instrument of sufficient sensitivity to be able to receive it.

In an attempt to illustrate exactly how big a task a systematic SETI program really is, consider the COSMIC HAYSTACK. We define this as the volume of multidimensional parameter space we might need to explore to find an extraterrestrial microwave signal. As noted, this volume is in fact nine dimensional!—three spatial dimensions, one temporal dimension, two possible senses of polarization between which the signal may be divided, an unknown frequency within the microwave region, an unknown modulation scheme for encoding any information content and, finally, an unknown strength for the transmitted signal. Because I cannot draw nine-dimensional figures, figure 20.1 is an attempt to represent the COSMIC HAYSTACK in three dimensions. To achieve this compression of dimensions, I have made

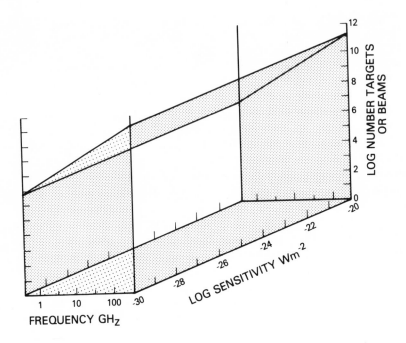

Fig. 20.1. The Cosmic Haystack. This is a three-dimensional representation of the nine-dimensional volume that would have to be searched for an extraterrestrial signal.

some assumptions that are consistent with the pragmatic limitations of conducting any SETI program with current technologies. First it is assumed that any signal is continuously present or at least has a relatively high duty cycle. Thus, when we look with the correct combination of the other eight parameters, it is likely that the signal will be "on" rather than "off." For example, THEY might choose to signal us by transmitting a single, spectacularly strong signal once only and never again. In that event we will probably not believe that we have received a signal, because we are unable to reconfirm or verify it. Indeed, one could argue that if THEY were so foolish as to adopt that transmission strategy, WE would not want to talk with THEM anyway! Therefore, we have chosen to neglect time in our depiction of the COSMIC HAYSTACK. Furthermore, we assume that every search will be conducted using instrumentation that is simultaneously sensitive to two orthogonal senses of polarization. In this way all of the power available in the signal will be collected and figure 20.1 can be thought of as being representative of either sense of polarization. In eliminating modulation as an axis for the figure, we must admit that we really cannot predict what type of signal an extraterrestrial engineer might choose to generate, but we can make some very educated guesses. If we further consider signals intended for detection, we can assert that whatever information encoding scheme is used, it will render the signal less noiselike rather than more so to the receiver, who cannot know in advance how to build a matched filter. Figure 20.1 has been drawn assuming that the signal can be detected and recognized without a priori knowledge of the modulation scheme. Decoding of any message is another question altogether. All these assumptions and assertions leave the three spatial dimensions plus frequency and signal strength to be represented.

The frequency axis of figure 20.1 runs from 300 megahertz (MHz) to 300 gigahertz (GHz) (that is, from 3×10^8 cycles/sec to 3×10^{11} cycles/sec). If we were attempting SETI observations from space, this would be the range of microwave frequencies over which the background noise from astrophysical sources is at a minimum. From the surface of our own planet, atmospheric oxygen and water vapor contribute substantially to the noise background at frequencies above 10 GHz. This additional noise will strain any ground-based SETI observations that we may undertake, but it should not be used to restrict the extent of the frequency axis of figure 20.1. Two of the three spatial dimensions can be combined onto one axis that depicts the number of different distinct directions on the sky at which we must point a radio telescope to complete the search. If we knew a priori the list of directions or objects from which a signal might possibly come, this axis would then represent the total number of "targets" on that list. However, it is possible to assert that every direction on the sky is potentially interesting.

Suppose, for example, that a very strong signal was being transmitted from the vicinity of a star just like our own Sun but located across the Galaxy from us. It is conceivable that our SETI instruments attached to a telescope might be capable of detecting this signal if we knew where to point the telescope. However, a solar-type star at that distance would appear far too faint to have been seen by an optical telescope and thus would never show up in any of our current catalogs of stars; we would never know that we ought to look there. To perform an all-sky survey, any telescope must tessellate the entire sky with beams, whose size is smaller for bigger telescopes and higher frequencies. For a given telescope the number of beams required to cover the sky is thus proportional to the collecting area of the telescope and to the square of the frequency of the observation. In figure 20.1 the slanting "roof" of the COSMIC HAYSTACK has been drawn assuming that the largest radio telescope currently in existence (Arecibo Observatory in Puerto Rico) could be used to make such a survey. For smaller telescopes the "roof" would be lower, but it would still slant because the telescope beam gets smaller as frequency increases.

The remaining axis in figure 20.1 labeled "sensitivity" combines the third spatial dimension (unknown distance to the signal transmitter) and the unknown power of the signal, and is measured in units of watts per square meter. Consider a 50 W light bulb; it is easy to see if you are standing 10 feet away from it, but it becomes invisible if you move back a mile or so. At any distance from the light bulb, a sphere of that radius constructed concentrically around the bulb will intercept all the photons given off. However, the number of photons intercepted by any single square centimeter on the surface of that sphere will decrease as the sphere gets bigger. After a certain distance the number of photons collected by the small area of your eye becomes so few that the eye cannot respond to them, and you can no longer see the light bulb. If the light bulb were 200 W instead, you could see it twice as far away. It is exactly the same with a radio telescope collecting photons over its surface area to feed to instruments that detect and record the signal. If the power of the signal transmitter increases, it is detectable at a greater distance. If the transmitter is farther away, then the number of photons collected by the telescope drops and eventually at some distance becomes so small that the signal detection instruments cannot respond to it. For an instrument to be able to detect a signal, the power collected from the signal must exceed the noise power internal to the instrument itself.

In the SETI program and all areas of astronomy, weak signals are the expected rule and therefore a great deal of effort is expended to make low-noise detectors. But how sensitive must our equipment be to detect a signal? The origin of this third axis has been assigned a value that is typical of the sensitivity achieved by radio astronomers when they survey the sky at interesting frequencies associated with radiation emitted by interstellar atoms or molecules. The implication is that if the signal transmitter is strong enough or close enough to imply this required sensitivity, then radio astronomers may well stumble upon it in the course of their work; a systematic SETI exploration should concentrate on the sensitivity regime that radio astronomers cannot reach. How good might we have to get? On Earth the single most powerful transmitter is the planetary radar attached to the radio telescope at Arecibo Observatory used to bounce echoes off the nearby planets to map their surfaces. If this transmitter emitted as intensely in all directions as it does in the narrow beam focused by the antenna, then it would be rated at 10^{13} W of effective radiated isotropic power. If this transmitter were located at the opposite end of the Galaxy from us, we could detect it with an instrument whose sensitivity was equal to the value at the outer limit of the sensitivity axis. We can, of course, hope that the ETIs have a technological capability in advance of our own and that they are somewhat closer than the far side of the Galaxy, but the limit chosen does reasonably represent the maximum effort we might need to invest. Having bounded the COSMIC HAYSTACK, it is reasonable to ask how much of it, if any, we have explored.

In the quarter century since Frank Drake's Project Ozma observations of the two nearby stars Tau Ceti and Epsilon Eridani, about three dozen SETI observations have been reported in the literature or media. In some cases a considerable number of hours of radio telescope time was involved. Does not that mean that we have already searched and THEY are not there? Not so. The problem is that until now such searches have been conducted with radio astronomy equipment intended for the detection of astrophysical sources. It is quite probable that any signal transmitted with the intention of having it recognized as an obviously artificial signal of intelligent origin or one intended for internal communication purposes will deliberately be generated in a way that makes it look very different from the signals the hardware was designed to detect, and the hardware may be completely insensitive to another signal type. Indeed, this is the outstanding problem with a SETI program: we don't know what it is we are looking for!

In figure 20.2, the parameter space explored by a few of the searches to date has been blocked out. This is a confusing figure, but the important point to note is that hardly any of the frequency axis has been touched, and those searches that cover a large number of directions on the sky have much poorer sensitivity than those with only a limited number of targets or beams on the sky.* This is the obvious consequence of the fact that radio astronomers build receivers for a limited set of frequencies, typically have available to them on order 100-1000 separate channels onto which they can sort incoming emission for more detailed analysis, and that they must compete for telescope time on instruments that are oversubscribed. Why should the number of channels for filtering data matter? When a natural source emits radiation, it does so over a range of frequencies,

---

*One way to visualize the degree to which the COSMIC HAYSTACK has been searched is to imagine a few sheets of paper hanging in a room. The search will have been completed when the room is full. Right now the room is virtually empty. (Eds.)

not just at a single frequency, but when you tune in your radio from
one station to another, your radio is searching for a signal that has
been transmitted on a single frequency only; the so-called carrier
wave.

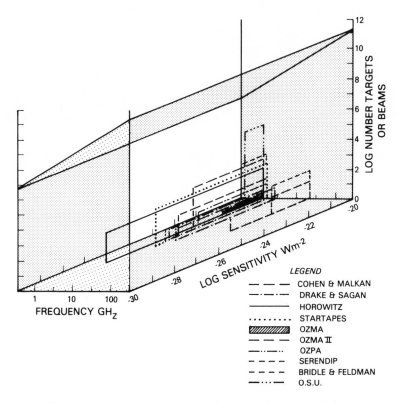

Fig. 20.2. Only a tiny fraction of the Cosmic Haystack has been
searched. Because searches done so far have been in narrow frequency
bands, one can imagine that the searched volume corresponds to a few
sheets of paper hung in a large and otherwise empty room.

For purposes of efficient communication, we do what Nature never
seems to do: we compress a lot of power into a single frequency. ETIs
might do so as well. Although we cannot know in advance the type of
the signal, a narrow band or monochromatic one is a good bet. So too
are pulses, in which a lot of power is compressed into a short period of
time. There are some types of pulses that Nature emits, but there are
others that would be suspected of being a result of intelligent origin.
Instrumentation being built for SETI is intended to be optimally
sensitive to at least these two types of signal and any other signal types

that can be accommodated with our current sophistication and hardware speed limitations on data processing. In particular, it is the narrow band signal that radio astronomy equipment may have been overlooking in the past. To be detectable a narrow band signal must deliver more power to the detector than the noise level within the detector itself. If the detector is divided into many narrow channels, then the noise present within any given channel will be much less than the noise within the entire detector, because this noise is proportional to the frequency width, but the strength of the narrowband signal will remain unchanged and reside within a single channel. A signal undetectable in a wide channel may become visible in a narrow channel owing to the increased signal-to-noise ratio. However, it takes a large number of narrow channels to explore a range of frequencies.

In figure 20.2 the most sensitive search depicted achieved a sensitivity to narrow band signal that came within a factor of 1,000 of the COSMIC HAYSTACK limiting value. This sensitivity was achieved by using an instrument with 128,000 channels, each of which was 0.015 Hz in width (compare with radio astronomy's 100 channels of width in excess of 1 kilohertz [kHz]). In all, this observational program explored only 2 kHz out of the nearly 300 million kHz of frequency in the COSMIC HAYSTACK. Yet this was not a foolish observational program, but rather one of the kind that guesses at certain "magic frequencies," at which it can be argued that ETIs are more likely to transmit because nature has defined them as a common point of reference for both sender and receiver. As the number of researchers entering such discussions grows, so does the number of proposed "magic frequencies," and in the end the uniqueness of these self-obvious frequencies vanishes. In sum total over the last quarter century, we have explored only a trivially small fraction of the COSMIC HAYSTACK; one part in 10^{17} to be precise. The situation will not improve without SETI-specific hardware.

A prototype of such hardware has been constructed as part of a NASA research and development program within the Life Sciences Office. In final form, this instrumentation will contain about 10 million individual channels, with widths as small as 1 Hz, and it will process data in real time at the rate of about 1 gigabit/sec with special purpose signal processors designed to be optimally sensitive to pulses and narrow-band carriers, while maintaining a degree of sensitivity to a wide class of more complex signal types. When used in a five-to-ten-year observing program on large radio telescopes worldwide and on the 34 m network of antennas of NASA's Deep Space Network of spacecraft tracking telescopes, it will be able to explore the volume of the COSMIC HAYSTACK sketched in figure 20.3. Although the current SETI observational program envisioned by NASA cannot explore the entire COSMIC HAYSTACK in any finite amount of time, the proposed search will represent an improvement in the volume of the parameter space covered by a factor of almost ten million over the combined efforts of the past twenty-five years.

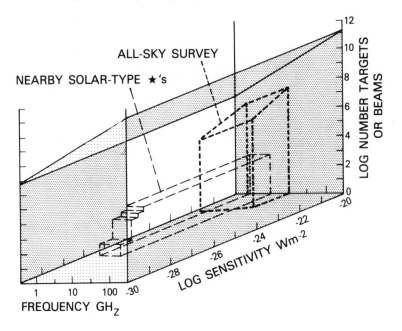

Fig. 20.3. Regions of the Cosmic Haystack that would be covered by proposed searches.

The bimodal search strategy depicted in figure 20.3 will include an all-sky survey conducted by driving the NASA 34 m antennas across the sky at a constant rate, but slowing down to increase sensitivity in directions of special interest to SETI and the astronomical community, such as the galactic plane. The frequency coverage will be from 1 to 10 GHz continuously and certain spot bands at higher frequencies as time and receiver availability permit. In this application the narrowest channel width will be 32 Hz, and the range of pulse periods and repetition rates allowed will be set by the short time during which the antenna beam looks at any direction on the sky. The sensitivity at 1 GHz should be about 10^{-23} W/m^2 and be poorer by a factor of 3 at 10 GHz. At the same time other instruments having channels as narrow as

1 Hz will be used to conduct a target search of the roughly 800 solar-type stars within 80 light-years of us over the frequency range 1 to 3 GHz. Using narrower channels and dwelling on each object for a longer time increases the sensitivity of these observations to 10^{-26} or 10^{-27} W/m² depending on the telescope employed. It is anticipated that the observing program can be completed with requests to various observatories for telescope time at a few percent of the available time for a few years.

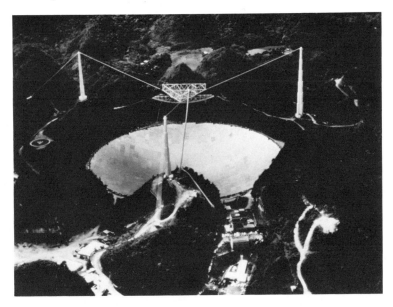

Fig. 20.4. The world's largest radio telescope is at the Arecibo Observatory in Puerto Rico. With the planetary radar attached, it could talk to a twin instrument 3,000 light years away. Arecibo Observatory is part of the National Astronomy and Ionosphere Center which is operated by Cornell University under contract with the National Science Foundation.

Might such searches be successful? It depends upon the number and strength of the transmitters out there. Neither search could detect the megawatt transmitters associated with our own TV broadcasting practices even at the distance to the nearest stars. However, the targeted search could detect a gigawatt transmitter (the power level of some defense radars or of a proposed orbiting solar power station) out to 30 light-years distance. The sky survey could detect the power level of the Arecibo planetary radar transmitter to a distance of 30 light-years, whereas the targeted search could detect such a transmitter near a solar type star out to a distance of 3,000 light-years. If there are more advanced technologies with the ability and inclination to transmit power at 100 times our current peak, then the sky survey could detect such transmitters out to a distance of 300 light-years, a volume that contains about 100,000 stars.

Colonization: Impact on Search Strategy

If colonization is indeed the galactic norm, then it argues that our immediate search strategies might profitably concentrate closer to home, because the nearest transmitter is likely to be far closer to us than is the nearest site of indigenous extraterrestrial intelligence. In particular, we must give serious consideration to cataloging our Solar System resources. Are asteroids and comets disappearing at an artificially high rate? Is there any other evidence of large-scale astroengineering? Clearly, a strong program of planetary exploration will begin to help here. In addition the very nearest stars become targets of particular interest under the colonization ground rules. Few people who debate the subject of extraterrestrial life would be so optimistic as to expect the abundance of life in the universe to have given rise to another example of intelligent species right next door. However, if colonization emplaced them there, then those colonists must be aware of our presence, and one scenario for an extraterrestrial signal becomes more realistic. It has been suggested that someone intercepting our early television transmissions (now 40 light-years away in all directions), which really means detecting the carrier, might choose to transpond that signal back to us, perhaps with slight modifications, as a method of attracting our attention. For each star and each TV channel combination, effort should be expended to monitor those frequencies on which a round trip could have been completed. Although the arrival of such a signal might be heralded by the outraged cries of viewers whose favorite broadcast was interrupted, it is more likely that the transponded signal will be extremely weak and require a much larger collecting area than a home aerial to receive. Such a strategy should properly be included as part of a systematic SETI observing program using large telescopes.

On a grander scale it could be argued that intercolony communication offers the best possibility for detectable electromagnetic signals. In this case arguments about the background noise level can no longer be used to infer the preferred portion of the spectrum. Colonies that are cognizant of one another's locations will be tempted to use tightly beamed communication signals, which can always be made detectable above the diffuse background of natural source radiation. In this case much higher frequencies and wide communication bandwidths may be the rule. Could we intercept such signals if we knew what frequency to try? A simple, model-dependent calculation indicates that there is something very much like a "confusion limit" in astronomy associated with this question. Suppose that the angular size of the tight communication beam is set by the desire to transmit from a colony near one star to a colony anywhere within the Solar System of a neighboring star; average interstellar distances of 4.5 light-years and average Solar System diameter of 10 AU imply a beam of 12 arcseconds. One would like to know how often Earth might fall into such a communication beam and thus be able to eavesdrop. This is equivalent to asking how often two colonies might line up and fall within a beam of 12-arc seconds diameter looking out from Earth. If we assume the extreme case that every star is colonized, then if one is looking in the galactic plane, on the average, it is necessary to go out a distance of 10,000 light-years before ensuring that two colonies will fall within any given beam. Looking out perpendicular to the plane, the stellar density drops too rapidly to achieve the average of two stars per beam area. Thus, if it were possible to conclude the probable frequency and transmitter power associated with such colonial conversation, it would only be necessary to plan an observational strategy around an instrument with sufficient sensitivity to detect the transmitter out to 10,000 light-years.

For now I shall stick to attempting to make a systematic exploration of the COSMIC HAYSTACK for the very first time, so that perhaps some limits can be placed on statement 4.

Acknowledgments

The author wishes to thank the organizers of the Interstellar Migration Conference for the invitation to participate in this unique discussion. Financial support was provided through NASA cooperative agreement NCC-2-36 with the University of California Space Sciences Lab.

References

Brin, G. D. 1983. The 'Great Silence': The Controversy Concerning Extraterrestrial Intelligent Life. *Q. J. Roy Astron. Soc.* 24: 283-309.

Hart, M. H., and B. Zuckerman eds. 1982. *Extraterrestrials: Where Are They?* Pergamon Press, New York.

EPILOGUE

EPILOGUE

Mankind is headed for the stars. That is our credo. Our descendants will one day live throughout the Solar System and eventually seek to colonize other star systems and possibly interstellar space itself. Immense problems—technical, economic, political, and social—will have to be solved for human life to spread through space. That almost goes without saying. Yet we do not think that, for example, the construction of closed ecosystems to sustain life in hostile space environments will prove utterly impossible, any more than devising viable space social systems will be. Nor do we think that the light-years that separate us from other stars will forever keep explorers (robot or human) and then settlers from leaving the Solar System. The resolution of these and other problems (including those we cannot now conceive) may prove enormously difficult and time-consuming on a scale of centuries or millennia. But we do think that ultimately they will be solved. The initiative to expand into space is already underway. Barring total nuclear war, a devastating collision with a comet or asteroid, or some other calamity on a worldwide scale, we think there is a good chance that this initiative will soon result in settlements in near space and that eventually our descendants will scatter among the stars.

We base our belief on an analysis of our evolutionary and historical past, as well as on an appreciation that *Homo sapiens* is a highly inquisitive, technologically innovating, expansionary animal. Yet we realize that our personal enthusiasm may cloud our judgment and that many critical questions about the whole notion of migrating into space can be raised. For example, one might ask if all this talk about space expansion isn't but an overblown extension of themes once dominant in Western society but now dated and discredited. More specifically: "Isn't the notion of space colonization just a projection of Western imperialism into a new and vastly larger realm?" Or: "Aren't all these elaborate schemes for transporting and supporting millions of people in space simply gross examples of technological thinking gone wild?" These and other similarly pointed questions posed to us before, during, and after the conference raise serious issues that demand some consideration. Although we cannot fully address all such issues in this epilogue, we would like to end this book with some indication of where our thinking lies.

The very term *space colony* does not go over well in the third world, and it is doubtful that substituting *space settlement* obviates the issue of unequal national participation in space. For citizens of nations only recently freed from foreign domination and perhaps still suffering from economic arrangements they deem neocolonial, the idea of space colonies or space settlements conjures up visions of advanced Western nations, particularly the United States, taking over vast chunks of real estate in space and then using these and their lead in space technology to further their economic power over other nations. Hence the appeal of the United Nations "Moon Treaty," which if it ever were ratified, would declare all resources in space to the "common heritage of mankind" and regulate economic exploitation.

From a short-term perspective this fear of being left out and left behind may be well founded, for only a few nations can now afford to go into space. Yet we would hasten to add that space is not the sole province of the Western capitalist nations that once so dominated the world. To be sure, the technological lead of the United States is impressive, particularly when augmented, as in the recent Spacelab mission aboard the Space Shuttle, by the rapidly developing skills of the European Space Agency. Yet socialist Russia remains a formidable space power, no matter how unsophisticated its space technology might be judged. The Russians have the most experience in long-duration manned spaceflight; and consonant with their Marxist faith in economic and technological progress, they appear more committed than any other nation to following a logically progressive program of space expansion up to and including actual colonization. Also, we should not forget that Japan has a vigorous space research and development program or that India, China, and a number of other third-world countries are passionately interested in space development. China even claims to be developing man-rated launchers and to be training their own astronauts. Here we cannot resist quoting from a letter from Joseph Needham, the noted historian of Chinese science and technology. In sending his regrets for not being able to attend the conference, he wrote: "My general feeling is that the Chinese of today would be just as keen to go exploring among the stars as any Westerners."

That China wants to shuck off its centuries-long slumber and reach out for the stars should alert us that just as motivation for expansion has waxed and waned among nations on Earth, so may new spacefaring nations rise as others fall. The current Russian-American hegemony in space cannot last forever. China, Japan, and other terrestrial nations will challenge it, and all those powers may eventually be surpassed by new and more vigorous space-based nations, which, like the Polynesian settlers of ocean space, will employ novel technologies and social organizations to expand the human frontier to the far reaches of the Solar System and beyond.

As to our contention that it is both inevitable and desirable that we strive to develop technologies to enable us to settle permanently in space, we reiterate that the use of technology to expand beyond Earth would be entirely consonant with the whole trend of human evolution. From the time those most adventuresome of apes left the tropical forest to seek a living in the grasslands of the African savanna, our ancestors have been inventing technology to adapt to new environments and to expand over the globe. There is a large techno-cultural distance between grubbing succulent roots from the soil of the savanna with digging sticks on the one hand and growing algae to provide both food and oxygen for Moon colonies, on the other. And it is a long way from sailing canoes to interstellar arks. But ever since our ancestors started using tools to survive and eventually flourish in new environments, the pattern of evolution by cultural as well as biological adaptation has been underway. Although the prospect of traveling and living in space might seem "unnatural" to many, it would represent a logical extension of the technological path our ancestors have been following for some 5 million years.

Yet we recognize that space will hardly be for everyone. Let us accept that early space settlements located, say, on the Moon or Mars or in orbit in the Earth-Moon system will be anything but Arcadian retreats. Instead, they are most likely to be cramped, austere, and hazardous outposts to which only the most adventuresome will apply (although the number of people who might volunteer even today might be surprising). Actually, not many people need to leave Earth to initiate the process of space migration or to keep it moving at the frontier. As the demographic studies of Birdsell and others show, small human groups can grow amazingly fast. Just as it probably took only a few bands of *Homo erectus* wandering north out of Africa—rather than a mass migration—to start the peopling of Europe and Asia, so too it may take only a tiny percentage of Earth dwellers, then Moon colonists, and so on, to forsake their birthplaces to move the population wave farther and farther outward.

Of course, to many people the idea of human communities living inside metallic cocoons may present a repellent, dehumanized vision of the future. But the stage of cramped orbiting habitats, pressure domes precariously perched on planetary surfaces, and the like, need not last forever; people should be able to learn how to further humanize space by building larger and more commodious structures (probably before the next century is very old) and eventually by terraforming inhospitable planets and other bodies. Still, the thought of applying more and more technology to make space habitable conjures up the spectre of further increasing the cultural uniformity and regimentation that is already growing apace on Earth as industrialization and the development of worldwide communication proceeds. Should billions someday live in myriads of intercommunicating space habitats orbiting the Sun, the development of a uniform high-technology space culture might seem regrettably inevitable. Yet precisely because space migration will not stop with the settlement of the inner Solar System, we do not fear such a development. Successful migration into far space will promote cultural diversity, not uniformity.

This is because if space technology really works and the dreams of settling space really do come true, then each isolated colony could in effect become a separate cultural experiment. First thousands, then millions of experiments in ways of living could flower in space and, as distances between colonies grow from Solar System dimensions to the light-years separating star systems, could come to fruition in the isolation of interstellar space where even radio messages might have to travel years before being heard. Compare that possibility with its polar opposite—an increasingly regimented and authoritarian one-world system on our overcrowded planet—and you will see the reason for our optimism about human society in space.

But this dispersion of humanity among the stars will bring much more than just refreshing cultural diversity. Eventually, it should lead to a development of a kind that few people other than science fiction writers have taken seriously: the appearance of new species, descended from humans but biologically distinct from us. In particular, if genetic engineering is combined with those "natural" forces (isolation, genetic drift, and harsh selective pressure) that will shape and accelerate evolution in space, a vision emerges of not one but many brave new worlds, each with a separate variety of human-descended but no longer *Homo sapiens* life. That may be hard to accept. Whether we consider ourselves to stand at the very pinnacle of the evolution of complex life or as having been created in God's image, it is sometimes difficult to imagine that we might simply be members of just another transient species in the continuing evolution and spread of intelligent life.

To pose a final crucial question: "Is it not but monumental *hubris* to plan (or even imagine) migration into space and the consequent creation of new forms of intelligent life?" The term hubris is of Greek origin and refers in classical ethical and religious thought to an overweening presumption that oversteps the limits placed by nature and the gods on human actions. It is a sin to which the main character in Greek tragedies are given. For example, in *The Persians* of Aeschylus, Xerxes arrogantly builds a bridge across the Hellespont, attempting to turn sea into land, an impious act for which he is punished by defeat at the hands of the Greeks at Salamis. Certainly, the history of global exploration and colonization is replete with many examples of hubris. For example, Captain James Cook, the greatest of the eighteenth-century explorers, accepted the homage of Hawaiians who worshiped him as the returning personification of their wandering god Lono, only to be killed a few weeks later by those same Hawaiians when he compounded his arrogance by attempting to take their ruling chief hostage. Then there is the case of Robert Falcon Scott, who ignored the need for the meticulous planning and for appropriate Norse-skiing and Eskimo dog-sledding technology that his rival Roald Amundsen was to use so successfully. Instead, Scott amateurishly raced for the South Pole relying primarily on what he considered to be the noble practice of "man-hauling" sledges; he paid for his presumption with his life and the lives of his companions. Many more examples of such individual hubris and the tragic consequences could be cited from the annals of exploration and colonization. But what is germane here is not the overweening presumption of individuals but the hubris of many people from many nations who dare to presume that not only will it be possible to colonize space but desirable as well.

Obviously, as two of those many people fascinated with the idea of space migration, we would not accept the notion that it is ordained by nature or God that humanity remain forever on Earth. We see no barriers in the near term and look forward to the next stages of space development. But we do not presume that *Homo sapiens* and descendant species will automatically and exponentially expand from Earth to rapidly fill the Galaxy and be its master. To presume that there are no unforeseen obstacles in the path of such a galactic expansion, including no other intelligent civilizations, might just turn out to be the most dangerous example of human hubris yet. Perhaps there are barriers, perhaps not; we just will not know until we have looked. That is why a concerted program of astronomical research must be an integral part of any expansionary strategy. This would include, along with standard astronomical observations and the use of robot spaceships to probe distant worlds, a vigorous search for extraterrestrial intelligence (SETI) by listening for radio signals and by any other feasible means. From this perspective SETI and other astronomical search activities would be more than pure research endeavors. They would be vital elements of the reconnaissance that must precede actual migration—in other words, the eyes and ears of the expansionary wave.

Should some unforeseen natural obstacles bar the course of expansion, or make it painfully slow and horrendously expensive, migration through the Galaxy might be severely restricted. Should we find that we are but one of many intelligent species in the Galaxy, expansion would similarly be checked, and an entirely new adventure of communication with and learning from others would begin. However, if we do find that we are truly alone and that there is nothing out there to materially impede expansion, then the whole Galaxy would be open to human, and posthuman, migration. If such a Galaxy-wide expansion does occur and if it could be observed from afar, it might well have the appearance of the regular advance of a population wave. However, if the experience in colonizing this one small planet is any guide, seen in detail this postulated galactic expansion would undoubtedly display its share of horror stories—of lost expeditions, failed colonies, conflict between competing branches of Earth-descended intelligent life, and so on. Then there is the prospect of some unfathomably sinister consequences of presuming that better, more space-adapted, intelligent life can and should be genetically engineered. Nonetheless, we do not think that the realization of these dangers—which may become more and more apparent as we begin to settle space—will necessarily deter expansion. Let us admit that we are a most hubristic species and that we are likely—unless our presumptuousness proves catastrophically fatal—to pass that trait on. Therefore, although we obviously cannot predict that human descendants will colonize the entire Galaxy, we are betting that they will try.

ABOUT THE AUTHORS

JOSEPH B. BIRDSELL, Emeritus Professor of Anthropology at the University of California, Los Angeles, was first trained in aeronautical engineering at the Massachusetts Institute of Technology where he received his S.B., but then switched to physical anthropology, receiving his Ph.D. in the subject from Harvard University. Subsequently, he has spent a lifetime studying the Australian Aborigines, their ecology, population structure, and microevolution. Dr. Birdsell's numerous publications on the Aborigines and similar groups contain information on small human populations that may be of great relevance to the first attempts to establish human populations in space.

GLEN DAVID BRIN is a postdoctoral Fellow at the California Space Institute, University of California at San Diego. He received his B.S. in Astronomy from the California Institute of Technology, a M.S. in Electrical Engineering and a Ph.D. in Applied Physics and Space Sciences from the University of California at San Diego. His principle professional interest is in the evolution and behavior of comets. Brin is also a science fiction writer of note, having received both the prestigious Hugo and Nebula Awards for his novel *Startide Rising*. He is currently working with T. Kuiper on a SETI book.

DAVID R. CRISWELL received his B.S. and M.S. in physics from North Texas State University, and his Ph.D. in space physics and astronomy from Rice University. After spending a decade with NASA's Lunar and

Planetary Institute, he has been working as a consultant with the California Space Institute (University of California), the Los Alamos National Laboratory, and several aerospace firms. Dr. Criswell's primary interest is in space industrialization, as evidenced by his many publications and patent applications in the field, as well as his public advocacy and his presidency of a Texas-based firm, Cis-Lunar Inc.

ALFRED W. CROSBY, trained in history at Harvard University where he received his A.B. and at Boston University where he received his Ph.D., is Professor of American Studies and Geography at the University of Texas, Austin. He focuses on environmental and medical history, and in particular on the consequences of the European exploration of and expansion over the globe. This is reflected in his *Columbian Exchange: Biological and Cultural Consequences of 1492*, published in 1972, and in a forthcoming book, *Ecological Imperialism: Europe Overseas, 1000-1900*.

BEN R. FINNEY is Professor of Anthropology at the University of Hawaii. He received his B.A. in Liberal Arts from the University of California, Berkeley, and his M.A. and Ph.D. from, respectively, the University of Hawaii and Harvard University. He has a long-standing interest in the settlement of the Pacific and the subsequent history of Pacific Island peoples, as reflected in such works as *Polynesian Peasants and Proletarians* (1973) and *Hokule'a: The Way to Tahiti* (1979). Now he is working primarily on the human implications of space exploration and colonization.

MICHAEL H. HART received his B.A. in mathematics from Cornell University, his law degree from the New York Law School, his M.S. in physics from Adelphi University, and his Ph.D. in astronomy from Princeton University. He has taught at the University of Maryland and Trinity University and held research positions at the Hale Observatories, the National Center for Atmospheric Research, and NASA's Goddard Space Flight Center. Although his major research field is planetary atmospheres, Dr. Hart has found time to write *The 100: A Ranking of the Most Influential Persons in History* (1978), as well as to explore the Fermi Question in a series of articles and in a book, co-edited with Ben Zuckerman, *Extraterrestrials: Where Are They?* (1982).

WILLIAM K. HARTMANN is Senior Scientist at the Planetary Science Institute in Tucson, Arizona. He received his B.S. in physics from

Pennsylvania State University, his M.S. in geology and his Ph.D. in astronomy from the University of Arizona. His major interests are in the origin of planets and the environmental and social effects of space exploration. Dr. Hartmann is noted for his skill at both writing about space subjects and illustrating them in paintings. His recent books include *Astronomy: the Cosmic Journey* (1982) and a co-authored work, *Out of the Cradle* (1984).

WILLIAM HODGES received his A.B. degree in economics from the University of California at Berkeley, where he is now working on a Ph.D. in the same subject. In addition to pursuing his studies in economics and demography, Mr. Hodges is working on a book on the foundations of logic and its application to quantum mechanics.

ERIC M. JONES, a Laboratory Fellow at Los Alamos, received his B.S. in Astronomy from the California Institute of Technology and a Ph.D. from the University of Wisconsin at Madison. He has done theoretical work on stellar explosions and on the effects of nuclear explosions. He served on the National Academy of Science Committee on the Atmospheric Effects of Nuclear Explosions (Nuclear Winter). He was the first to publish models of interstellar migration, and his popular lecture "Interstellar Migration and the Human Experience" has been given to such groups as the L-5 Society, the Chicago Society for Space Studies, Sigma Xi, and the NASA Lunar Base Working Group.

RICHARD B. LEE, Professor of Anthropology at the University of Toronto and President of the Canadian Ethnology Society, received his B.A. and Ph.D. in anthropology from, respectively, the University of Toronto and the University of California, Berkeley. His major research focus has been on hunter-gatherer life, especially in Africa, and its implications for social evolution, and has published a number of books on the subject including *Kalahari Hunter-Gatherers* (with I. DeVore, 1976) and *The Dobe !Kung* (1984). In 1971 Dr. Lee was one of two anthropologists who participated in the historic meeting on Communication with Extraterrestrial Intelligence held in Soviet Armenia. He is currently working on a book about the origin of early state societies.

WILLIAM I. NEWMAN is Associate Professor of Earth and Space Sciences and Astronomy at the University of California, Los Angeles. He received his B.Sc. and M.Sc. in physics from the University of Alberta, and his M.S. and Ph.D. in astronomy and space sciences from Cornell University. Dr. Newman's interests range from mathematical biology and geophysical dynamics to high energy astrophysics and galactic dynamics, and has recently published in these fields in such journals as the *Journal of Theoretical Biology, Geophysical Research Letter, Astrophysics Journal,* and *Icarus.*

EDWARD REGIS JR., Associate Professor of Philosophy at Howard University, received his B.A. from Hunter College, and his M.A. and Ph.D. in philosophy from New York University. When not pursuing topics in academic philosophy, Dr. Regis critically examines issues in space development, the search for extraterrestrial intelligence, and other areas seldom visited by philosophers. He is the editor of *Gewirth's Ethical Rationalism* (1984), and a forthcoming book, *Extraterrestrials: Science and Alien Intelligence.*

CARL SAGAN is David Duncan Professor of Space Sciences and Director of the Laboratory for Planetary Sciences at Cornell University. He was educated at the University of Chicago, where he received his A.B. degree, his S.B. and S.M. degrees in physics and his Ph.D. in astronomy and astrophysics, and has subsequently received numerous honorary degrees and awards. Dr. Sagan, who is noted for both his scholarly and popular works in astronomy and other fields, is a leader in the exploration of the Solar System and the search for extraterrestrial intelligence as well.

DOUGLAS W. SCHWARTZ is President of the School of American Research, a center for advanced studies in anthropology located in Santa Fe, less than an hour's drive from Los Alamos. He received his B.A. and Ph.D. in anthropology from, respectively, the University of Kentucky and Yale University, and he was awarded a Litt.D. by the University of New Mexico in 1981. Dr. Schwartz has conducted extensive archaeological excavations in the southwestern and the southeast United States, and has published numerous monographs on these investigations. He has also conducted a social anthropology study of a community on the island of Stromboli, Italy.

JILL C. TARTAR received her Bachelor in engineering physics from Cornell University, and earned her M.A. and Ph.D. in astronomy from the University of California, Berkeley. She is Associate Research Astronomer in the astronomy department at the University of California, Berkeley, and is currently under contract to NASA's Ames Research Center where she works on the search for extraterrestrial intelligence. Dr. Tartar has published and lectured worldwide on SETI as well as more conventional astrophysical topics, and is now writing a textbook on SETI with Frank Drake.

NANCY MAKEPEACE TANNER is Associate Professor of Anthropology at the University of California at Santa Cruz. She received a B.A. in Pre-Med from the University of Chicago and a Ph.D. in anthropology from the University of California at Berkeley. She has worked in human evolution, and championing both the importance of gathering by females in early human development and the utility of a chimpanzee model in understanding early hominid evolution in cultural anthropology, through her work with the Minangkabau. Her work in human evolution appears in her book *On Becoming Human*.

JAMES W. VALENTINE holds a joint appointment as Professor of Geological Sciences and Biological Sciences at the University of California, Santa Barbara. He received his B.A. in geology from Philips University, and his M.A. and Ph.D. in the same subject from the University of California, Los Angeles. Dr. Valentine's major fields of research are paleoecology and macroevolution. He is the author of *Evolutionary Paleoecology of the Marine Biosphere* (1973), the editor of a forthcoming book, *Phanerozoic Diversity Patterns: Profiles in Macroevolution*, and is working on two more books on evolution.

KENNETH W. WACHTER is Associate Professor of Demography and Statistics at the University of California, Berkeley. He received his B.A. in history and literature from Harvard University, his M.A. in mathematics from Oxford University, and his Ph.D. in statistics from Cambridge University. Dr. Wachter's interests range from mathematical demography and statistics in historical studies, to computer simulation and multivariate analysis. His enthusiasm for applying quantitative analysis to societal questions is reflected in his book, *Statistical Studies of Historical Social Structure* (1978), and a forthcoming work, *A Mathematical View of Human Prehistory*.

INDEX